Astonishment and Science

VERITAS
Series Introduction

"... the truth will set you free" (John 8:32)

In much contemporary discourse, Pilate's question has been taken to mark the absolute boundary of human thought. Beyond this boundary, it is often suggested, is an intellectual hinterland into which we must not venture. This terrain is an agnosticism of thought: because truth cannot be possessed, it must not be spoken. Thus, it is argued that the defenders of "truth" in our day are often traffickers in ideology, merchants of counterfeits, or anti-liberal. They are, because it is somewhat taken for granted that Nietzsche's word is final: truth is the domain of tyranny.

Is this indeed the case, or might another vision of truth offer itself? The ancient Greeks named the love of wisdom as *philia*, or friendship. The one who would become wise, they argued, would be a "friend of truth." For both philosophy and theology might be conceived as schools in the friendship of truth, as a kind of relation. For like friendship, truth is as much discovered as it is made. If truth is then so elusive, if its domain is *terra incognita*, perhaps this is because it arrives to us—unannounced—as gift, as a person, and not some thing.

The aim of the Veritas book series is to publish incisive and original current scholarly work that inhabits "the between" and "the beyond" of theology and philosophy. These volumes will all share a common aspiration to transcend the institutional divorce in which these two disciplines often find themselves, and to engage questions of pressing concern to both philosophers and theologians in such a way as to reinvigorate both disciplines with a kind of interdisciplinary desire, often so absent in contemporary academe. In a word, these volumes represent collective efforts in the befriending of truth, doing so beyond the simulacra of pretend tolerance, the violent, yet insipid reasoning of liberalism that asks with Pilate, "What is truth?"—expecting a consensus of non-commitment; one that encourages the commodification of the mind, now sedated by the civil service of career, ministered by the frightened patrons of position.

The series will therefore consist of two "wings": (1) original monographs; and (2) essay collections on a range of topics in theology and philosophy. The latter will principally be the products of the annual conferences of the Centre of Theology and Philosophy (www.theologyphilosophycentre .co.uk).

Conor Cunningham and Eric Austin Lee, *Series editors*

---

1. Note: Nathan Kerr, *Christ, History, and Apocalyptic*, although volume 3 of the original SCM Veritas series, is available from Cascade as part of the Theopolitical Visions series.

Astonishment and Science

Engagements with William Desmond

Edited by

Paul Tyson

Contributors

William Desmond   Isidoros Katsos
Jeffrey Bishop   Steven Knepper
Spike Bucklow   Simone Kotva
Richard J. Colledge   Simon Oliver
Andrew Davison   D. C. Schindler
Michael Hanby   Paul Tyson
Jonathan Horton

CASCADE *Books* · Eugene, Oregon

ASTONISHMENT AND SCIENCE
Engagements with William Desmond

Veritas

Cascade Books
An Imprint of Wipf and Stock Publishers
199 W. 8th Ave., Suite 3
Eugene, OR 97401

www.wipfandstock.com

PAPERBACK ISBN: 978-1-6667-3340-2
HARDCOVER ISBN: 978-1-6667-2805-7
EBOOK ISBN: 978-1-6667-2807-1

Cataloguing-in-Publication data:

Names: Tyson, Paul [editor]. | Desmond, William, 1951– [author]

Title: Astonishment and science : engagements with William Desmond / edited by Paul Tyson.

Description: Eugene, OR: Cascade Books, 2023 | Series: Veritas | Includes bibliographical references.

Identifiers: ISBN 978-1-6667-3340-2 (paperback) | ISBN 978-1-6667-2805-7 (hardcover) | ISBN 978-1-6667-2807-1 (ebook)

Subjects: LCSH: Desmond, William, 1951– | Science—Philosophy | Religion and science | Metaphysics | Philosophy—Metaphysics

Classification: BD111 T97 2023 (print) | BD111 (ebook)

We acknowledge with thanks the permission granted by the Catholic University of America Press to reproduce chapter three from William Desmond's book *The Voiding of Being* (2020) as chapter 1 in this book.

Contents

Illustration

The Toplou icon, *Megas ei Kurie* (Great are You, O Lord), 1770 | 170
Painted on wood by Ioannis Kornaros.
Image source:
Wikimedia Commons
https://commons.wikimedia.org/wiki/File:Megas_Ei_Kyrie
    _from_Ioannis_Kornaros_1770.jpg
Accessed: 25 May 2022

Contributors

Jeffrey P. Bishop is professor of philosophy and professor of theological studies at Saint Louis University, where he holds the Tenet Endowed Chair in Bioethics. He is the author of *The Anticipatory Corpse: Medicine, Power, and the Care of the Dying* (University of Notre Dame Press, 2011) and the co-author (with M. Therese Lysaught and Andrew A. Michel) of the Expanded Reason Award-winning book, *Biopolitics After Neuroscience: Morality and the Economy of Virtue* (Bloomsbury, 2022). Bishop's recent work has returned to the intuitive methodology of Henri Bergson and the French spiritualist tradition in philosophy. His most recent book project explores the relationship between structure and dynamism in living beings in relation to the good of living beings, to develop a philosophical anthropology based in the activity of being human, especially focusing on the relationship of the tool-making and myth-making activities of the human animal.

Spike Bucklow is a scientist and professor of material culture at the University of Cambridge where he had the privilege of working on England's oldest and best preserved medieval altarpieces (the *Westminster* and *Thornham Parva* retables). These experiences alerted him to the dangers of applying modern worldviews to historic cultures and, attempting to avoid those dangers, he taught himself pre-modern natural philosophy as an operative system. The resulting insights have led to a number of books, including *Children of Mercury* (Reaktion, 2022).

Richard J. Colledge is associate professor of philosophy, and associate dean in the Faculty of Theology and Philosophy, at Australian Catholic University, based on its Brisbane campus. He has various published papers in the philosophy of religion, the history of metaphysics, and the traditions of phenomenology and hermeneutics. He is currently working on a book concerned with Martin Heidegger's abyssal realism.

Andrew Davison is the starbridge associate professor in theology and natural sciences, University of Cambridge, working at the intersection of theology, science, and philosophy. He holds undergraduate degrees and doctorates in both natural science and theology. He has recently worked on astrobiology (*Astrobiology and Christian Doctrine*, Cambridge University Press, 2022) and has an abiding interest in the "metaphysics of participation" as a structuring principle in Christian theology (*Participation in God: A Study in Christian Doctrine and Metaphysics*, Cambridge University Press, 2019). He writes and lectures widely on theology for a public audience.

William Desmond is David Cook Chair in Philosophy at Villanova University, and professor of philosophy emeritus at the Institute of Philosophy, Katholieke Universiteit Leuven. He is also the Thomas A. F. Kelly Visiting Chair in Philosophy, at Maynooth University, National University of Ireland. His work is primarily in metaphysics, ethics, aesthetics, and the philosophy of religion. He is the author of many books, including the trilogy *Being and the Between* (1995), *Ethics and the Between* (2001), and *God and the Between* (2008). Most recently he published *Godsends: From Default Atheism to the Surprise of Revelation* (2021). He is past president of the Hegel Society of America, the Metaphysical Society of America, and the American Catholic Philosophical Association. He is the recipient of Doctor of Literature (DLitt), *honoris causa* from Maynooth University.

Michael Hanby is associate professor of religion and philosophy of science, Pontifical John Paul II Institute, Washington, DC. Hanby is author of *No God, No Science?* (2013), which reassesses the relationship between the doctrine of creation, Darwinian evolutionary biology, and science more generally. He is also author of *Augustine and Modernity* (2003), which is simultaneously a re-reading of Augustine's trinitarian theology and a protest against the contemporary argument for continuity between Augustine and Descartes.

Jonathan Horton is an Australian-based practising King's Counsel (KC). He holds a PhD (law) from Edinburgh University in legal history and the philosophy of law. His interest is public law and constitutional law, and understanding the nature of legislative proliferation and its underlying influences. Jonathan's work explores, as part of this, repositories of law outside the "made" and enacted.

Isidoros C. Katsos is a British Academy postdoctoral fellow at the Faculty of Theology and Religion at the University of Oxford and a junior research fellow at Campion Hall. He also holds academic positions in the University of Cambridge, at the Von-Hügel Institute and the Centre for the Study of Platonism, and he was previously a postdoctoral fellow at the Hebrew

University of Jerusalem. He holds a PhD in philosophy of religion and patristics from the University of Cambridge, under the supervision of Rowan Williams; and a PhD in human rights law from the Free University of Berlin. His academic interests lie in all aspects of early Christian philosophy, informed by contemporary discussions. His has published *The Metaphysics of Light in the Hexaemeral Literature: From Philo of Alexandria to Ambrose of Milan* (Oxford University Press, 2022). He is currently working on a multi-annual project on *Christian Human Rights: Ancient and Postmodern*. Fr. Isidoros is a Greek-Orthodox priest bearing the rank of archimandrite.

Steven Knepper is associate professor of English at Virginia Military Institute. He is the author of *Wonder Strikes: Approaching Aesthetics and Literature with William Desmond* (SUNY, 2022), and he is currently editing the essay collection *A Heart of Flesh: William Desmond and the Bible* (Cascade Books, forthcoming). He is also co-writing a critical introduction to the philosophy of Byung-Chul Han.

Simone Kotva is research fellow at the Faculty of Theology, University of Oslo, and affiliated lecturer at the Faculty of Divinity, University of Oslo. She is the author *of Effort and Grace: On the Spiritual Exercise of Philosophy* (Bloomsbury, 2020), the first book in English to situate Simone Weil's thought within the history of French reflexive or spiritualist philosophy, from Maine de Biran to Henri Bergson and Alain (Émile Chartier). Simone has published widely on philosophy as a spiritual exercise and way of life, and her work is a critical intervention seeking to rethink the legacy of Pierre Hadot and Michel Foucault by engaging feminist, decolonial, and ecological theologies of practice. She is currently completing a book on agency and passiveness, *Ecologies of Ecstasy: Practicing Philosophy through Mystical and Vegetal Being*.

Simon Oliver is Van Mildert Professor of Divinity at Durham University and residentiary canon of Durham Cathedral. His teaching and research lie in the fields of Christian doctrine and philosophy, particularly the Thomist tradition and the doctrine of creation. He is author of *Philosophy, God and Motion* (2013) and *Creation: A Guide for the Perplexed* (2017). His forthcoming book is entitled *Creation's Ends: Teleology, Ethics and the Natural*. He is currently engaged in a research project on theology and phenomenology's enquiry into the nature of life.

D. C. Schindler is professor of metaphysics and anthropology at The John Paul II Institute at The Catholic University of America in Washington, DC. He has focused his work on the nature of the transcendentals in metaphysics

(beauty, truth, goodness) and on the fundamental anthropological correlates of these: love, reason, and freedom. The author of ten books—including *Plato's Critique of Impure Reason: On Truth and Goodness in the Republic*
(CUA, 2008), *Love and the Postmodern Predicament: Rediscovering the Real
in Beauty, Truth, and Goodness* (Cascade, 2018), and *The Politics of the
Real: The Church between Liberalism and Integralism* (New Polity, 2021)—
Schindler is also a translator of works in French and German and one of the
editors of the North American edition of *Communio: International Catholic
Review.*

Paul Tyson writes about Christian Platonism and theological metaphysics
and epistemology, the theology of science, theological sociology, the sociology of knowledge, and the theology and politics of money. His recent books
include *Seven Brief Lesson on Magic* (Cascade, 2019), *Kierkegaard's Theological Sociology* (Cascade, 2019), *Theology and Climate Change* (Routledge,
2021), and *A Christian Theology of Science* (Baker Academic, 2022).

Introduction

Paul Tyson

SCIENCE IS THE NATURAL philosophy of Western modernity.[1] As a natural philosophy, modern science has gradually come to perform important cultural functions that used to be done by theology and metaphysics. By now—in our broadly materialist and secular public-knowledge environment—our shared tacit cultural understandings of the nature of reality, the meaning of our own humanity, and the contours of cosmic purpose and moral truth (usually understood in terms of their convenient absence), are deeply shaped by our natural philosophy. In a rather historically unique manner, our natural philosophy now firmly molds what sociologists of knowledge call our social reality, to the exclusion of traditional metaphysics and theology from the domain of public knowledge.[2] In this context, thinking about where and how science has become a theology and metaphysics excluding

1. Technically, it is more philosophically accurate to say that the methods, broadly accepted facts and theories, and the epistemic habits, technological uses, and institutional authority of modern science cannot be meaningfully extracted from the assumed natural philosophy of Western modernity. So *technically*, modern science (as a multi-faceted body of knowledge and a set of knowledge-concerned practices) can be held as conceptually distinct from modernity's assumed natural philosophy (broadly shared beliefs about the nature and meaning of nature). Sociologically, however, our distinctly modern understanding of scientific truth is so intimately entailed in the way we collectively see nature, supernature, humanity, and the realms of public facts, private beliefs, instrumental power, and realistic action that their technical distinction looks pretty academic in reality. In terms of what sociologists of knowledge call our social reality, science really *is* our natural philosophy.

2. See Harrison, *The Territories of Science and Religion* for a careful look at the manner in which there was a historical tipping point in the late nineteenth century when the territories of science and religion were firmly demarcated, with the public domain of knowledge given over to science, and personal domains of meaning and faith given over to religion.

public-domain ideology—"scientism"—is an important concern for anyone still interested in traditional metaphysics and theology.

But perhaps I am speaking in too shrill a manner. Many Christians and people of other faiths who are scientists have found the nineteenth-century separation of science from religion quite serviceable. Science tells us the "what" of observable reality, religion (and perhaps metaphysics) tell us the "why." The modern world holds that facts and quantities are entirely different knowledge and understanding (belief?) categories to meanings and qualities; there is no inherent reason why they should clash. Provided science keeps to the factual and theoretical domain of "what" and faith and philosophy keep to the meaning and value domain of "why," the territorial arrangement can be, and should be, mutually constructive.

The nature of the relationships between the practices and knowledge-constructions of modern science, the revelations and community-forma-tions of religion, the contemplations of metaphysical explorations, and the distinctive configuration of the modern categories of public knowledge and private meaning are genuinely complex. Such a dynamic and multi-faceted matrix requires a high degree of intellectual finesse across a complex field of intersecting modalities of knowledge, understanding, belief, and power to do it justice. Few thinkers in our times of high specialization have really got what it takes to perform such an endeavor well. But there is a very capable thinker, a truly able metaphysical theologian, who we can call on to help us in this very important matter: William Desmond.

◆ ◆ ◆

William Desmond has a distinctive voice. Some hear that voice as clear as a bell and as piercing as truth. Others simply cannot follow the meaning his seemingly mysterious words convey.

In preparing this book, I was discussing with D. C. Schindler[3] the striking manner in which some people "get" William, and others do not. D. C. Schindler, like myself, finds it hard to understand how anyone might not be moved to lyricism by the beautiful clarity and profound insights of this astonishing bard of metaphysical theology. We were trying to work out what the "problem" might be, for we both know some super-intelligent and profound thinkers who—one might say—are tone deaf to William's songs. We thought perhaps "the problem" has something to do with a key concept in William's thought—the "in between."

---

3. I would ordinarily refer to D. C. Schindler as David, excepting he often finds people confuse him with his famous theologian father, David L. Schindler. So, in this introduction, the younger David Schindler is referred as D. C. Schindler.

When a knowing subject—a thinking, feeling, perceiving person—interacts with another person or thing, the magic happens "in between" the knower and the known. Among other things, the subject is revealed to themselves in knowing the object, and the object likewise is engaged and changed by being known. When two personal beings know each other, the space "in between" them, where they meet, is where the Spirit moves. Life happens *between* beings. And it cannot be bottled. What is most important about both the knower and the known is beyond capture within either of them. So one has to be able to play outside of the (impossible) subjective certainties of total knowledge capture within us as knowers, and outside of the (impossible) mastery of the outer world—the world of beings as objects of knowledge—to follow William's songs of thought. But this goes against the grain of how we are trained in knowledge-mastery and instrumental knowledge-construction. And perhaps those who are most advanced among us in the training of our knowledge culture are most troubled by William's songs. For his songs are always in the key of love, rather than the key of mastery. This does not mean William is less than expert concerning what *can* be said about knowers and the known, but clarity on that front is the leaping-off point into the un-masterable joy of what is really important—the "in between," the playground of the Spirit. One might say William has a pneumatological metaphysics of the natural world, but our knowledge culture is increasingly (mis-)constructed by a functionally materialist metaphysics of the (counterfeit) spiritual world. Hence, communication difficulties between William and our dominant knowledge-culture are entirely to be expected. And that will happen most strikingly when we come to talk about modern science.

♦ ♦ ♦

Part One of this book is a chapter from one of William's books, reproduced here with the kind permission of the original publisher.[4] The chapter is titled "The Dearth of Astonishment: On Curiosity, Scientism, and Thinking as Negativity."[5] William's chapter can be read on its own as an exploration of the troubled relationship between modern science and what he identifies as the three modalities of wonder: astonishment, perplexity, and curiosity. For anyone interested in the nature of modern science and its distinctive relationship with one mode of wonder (curiosity), and the particular ways in which modern science often seems to fail to relate in a satisfactory manner to other modes of wonder (perplexity and astonishment), this is a very

---

4. Many thanks to the Catholic University of America Press.
5. This is chapter 3 in Desmond, *The Voiding of Being*, 96–125 [15–42].

helpful and important read. For the distinctive signatures of scientism that arise from its functionally and materially focused curiosity, and the manner in which a scientistic knowledge of "the real world" of functional material operations becomes isolated from wisdom, which is metaphysically and theologically open to the higher modalities of wonder, is integral with the distinctive forms of spiritual poverty native to knowledge, power, and meaning in the life-world in which we now live.

William reminds us that astonishment is an open, "in between," Spirit-blown experience, marked by a complex relation to that which exceeds the grasp of modern scientific knowledge, and that a meaningful understanding of science itself cannot be produced from within the determinate and instrumentally masterable categories of modern scientific curiosity. This is a difficult message to hear if we are experts in any of the domains of modern determinate knowledge where reduction, specialization, quantifiable demonstration, and the sifting and discarding of the unknowable from the knowable are integral to our approach to publicly valid truth claims. Yet, to all the thinkers engaging with William's essay in this volume, the licence to include the modalities of astonishment and perplexity in how we approach public truth and the realm of factual and instrumental knowledge is nothing short of an intellectual and spiritual liberation.

◆ ◆ ◆

Part Two of this book is a collection of engagements with William's very important chapter on the three modalities of wonder. As mentioned, the contributors to Part Two have all found William's thoughts on astonishment and science very helpful. But this is a very challenging arena because the openness of metaphysical theology and the defined and contained knowledge of modern science do not often get in the same room, let alone try and speak to each other. So it seems that the people most needing to hear what William has to contribute—scientists, and people who think of themselves as comfortably at home in a science-defined life-world—might be less naturally able to hear him than the people who normally read and enjoy William's work.

One key aim of this book is to build a bridge to William's chapter for people whose world is deeply shaped by modern science, who we think would be drawn to William's insights if they could just understand enough of the dialect and tonality of his song for it to make sense to them. The way we have approached this is to collect a range of thinkers to write short pieces on either how astonishment engages their own world of thought, understanding, and practise, or to write pieces that engage directly with William's chapter. These chapters are wonderful works of insight in their own right, but they are not—*per impossibile*—written in William's voice. Sticking with a musical metaphor, these engagements play variations on William's astonishment-and-science theme in different idioms and ensembles to William's orchestral suite, which might help tune the ear to his thinking. So whilst the first section of this book opens with William's essay, if you find it initially hard to follow, perhaps move to Part Two first, reading engagements with William's work, and then go back to William's opening chapter. The way in which different aspects of William's insights are taken up by different engagements with his chapter, allows for more ways into William's thought world, at least one of which is likely to connect well with any reader.

A further aim of this book is scholarly engagement with William's work, in the more traditional register. So there will be some questioning, some application in areas not directly addressed by William, and some hard-headed analysis of his argument. William responds to all contributors' engagements in Part Three of this book.

• • •

Here is a short introduction to each of the twelve engagements with William's opening essay:

Chapter 2 is by Spike Bucklow. Spike's PhD is in chemistry and he works in Cambridge in art preservation and restoration. This job puts him in the zone of modern material sciences, requires an intimate knowledge of old craft practices, and draws on Spike's empathetic understanding of the (usually religious) objects of higher meaning that medieval art and its material engagements with the natural world opens up. Unsurprisingly, Spike has a deep understanding of the worldview and craft practices of the Middle Ages. This is a sympathetic cosmos were the natural elements have agency, where craft gilds foster contemplative manual disciples to learn the manners of our tacit understandings and co-agent relationships with the material world, and where nature itself is a sacrament. What we now call "science" was unknown in the Middle Ages, in large measure because natural knowledge was embedded in an astonishing openness to the metaphysical and the theological in the natural. Familiarity with this world gives Spike a remarkable perspective—by way of contrast—on the way we now think about "science." His chapter is a delight to read as he unpacks the surprising intimacy and richness of old craft skills and their relation to higher meanings and the natural world, both in the pre-modern world he preserves as a job, and in his own practice of modern science in doing that preserving.

In the third chapter, Steven Knepper opens up astonishment in a very direct and beautiful manner, via childhood. Steve is by nature first and foremost, a poet. The world strikes his being powerfully and he responds, richly, in carefully built words, ones bearing openness to the astonishment of being itself. Drawing on the dazzling spiritual luminosity of a father's relation with his young child, Steve also probes his own childhood experiences of astonishment in the natural world. He relates this back to William's chapter in a most helpful way. This is a pearl of careful and insightful writing.

Simone Kotva's contribution in chapter 4 richly engages the work of William Desmond with another famous Belgian thinker, Isabelle Stengers. Stengers is one of the premier continental philosophers of science in our day, and she comes to that field from chemistry, eco-feminism, and what Simone insightfully describes as "mystico-pragmatism." Following on from Steven Knepper's interest in how a child responds to astonishment, Simone unpacks the hesitance of the child as expounded in Stenger's "slow science" and relates this back to Desmond's thinking. Lingering with astonishment, and being open—quietly, patiently, receptively—to the marvel of the real, allows the world around us to speak its own language. The way modern science too often rushes headlong to master and control reality speaks not of wonder, but of its overcoming via the will to epistemic mastery. In critiquing this will to mastery, Desmond's philosophical theology and Stengers' not-theological philosophy of science connect with each other in very fruitful ways. Simone's chapter convincingly draws on Simone Weil's insights in order to connect Stengers' philosophy of science with Desmond's theological metaphysics. This is a theoretically demanding chapter—not surprisingly, as Simone is one high powered thinker—but those with an interest in the philosophy of science will find it most rewarding.

My own contribution to this volume, in chapter 5, looks at how a Desmondean outlook on astonishment shapes my approach to the social sciences. As a sociologist I appreciate that having a scientific approach to understanding certain measurable and seemingly "mechanical" aspects of human societies can be illuminating, provided one bears in mind what such an approach can and cannot achieve. Too often, however, the way in which we unwittingly make false assumptions about what science itself is, involves failing to see what our social scientific perspectives cannot disclose. For this reason, trying to re-think what a social scientific perspective is, from the grounds of metaphysical theology, would produce a better (and perhaps profoundly different) sort of social science. For what it means to be human, and what it means for humans to live in communities, are astonishing and darkly luminous mysteries. If our metaphysics of human reality is inadequate, our social scientific data and theories about the functions, practices, meanings, and measurable signatures of human life will be inadequate. This chapter, then, is an attempt to apply William's theorizing about astonishment to the social sciences.

Andrew Davison, who writes chapter 6, is one of those rare and remarkable dual citizen types with PhDs in both biochemistry and theology. In responding to William's chapter, Andrew explores the question of whether modern scientific curiosity is only dangerous, or whether it has a good side as well. This is much the same question as to whether science can be validly separated from scientism or not (a point of discussion that arises in a number of chapters in this volume). By interfacing the sociologist T. S. Kuhn with William's chapter, Andrew argues that William allows for a valid role for modern curiosity (and hence modern science) at the same time that Kuhn needs Desmond's metaphysics to allow science to at least aspire to some form of genuine realism. This is a beautiful and intriguing chapter, very clearly argued, and it draws on Andrew's inside knowledge of the community of practice that is the context in which modern science is actually done.

Chapter 7 is by Richard Colledge, a philosopher of rare type and ability. His chapter beautifully unpacks the manner in which William's chapter draws on and engages with Heidegger, as regards the relationship of science and technology to wonder and being. This chapter also helps explain—incidentally—why William is often met with perplexity in the Anglo-American scene; English language philosophy is often a discourse of thought that really does not understand a lot of what is going on in continental thought, even though we also live under the shadow of Kant and Hegel. Richard's chapter is very helpful in showing Anglophone thinking that William is not simply making up his own terms and metaphysical categories, but is engaged in the world of thought that is (to be frank) more metaphysically and theologically advanced than the often explicitly linguistically and scientistically reductive landscape of Anglophone philosophy. In philosophy, the continentals have been less metaphysically closed down by the marvels of the modern scientific age than have the Anglophones, and part of William's "problem" is that he is a continental theological metaphysician writing (often) in English.

Chapter 8 is by Jeff Bishop, who is an ethicist, philosopher, theologian, and medical doctor. This is an absorbing chapter. If you want to know what a concrete application of William's understanding of the various modalities of wonder are, this chapter will give it to you. Most professions—such as medicine and law—deal in the domain of determinate knowledge. That is, the body of knowledge that is medical science, and the body of knowledge that is legal interpretation, must be—to some extent—mastered, by the proficient doctor or lawyer. But, perhaps unlike for the pure scientist who is simply trying to crack open one specific door at the edge of present knowledge, the doctor and the lawyer must always relate their knowledge to profound mysteries that transcend our determinate knowledge; the living reality of the patient and the nature of justice. Doctors and lawyers can, of course, fail to relate their determinate knowledge appropriately to the mysteries that are inextricable from the practice of medicine and law. Yet excellent doctors and lawyers are those who develop a skilled proficiency in the complex art of marrying the determinate knowledge mastery and technical skills of their profession with an astonished reverence for the people and transcendent horizons that they serve. Jeff unpacks his own journey on that maturing to show us that the tendency for science to move from astonishment to perplexity to curiosity and to mastered knowledge goes in the reverse direction for the practice of medicine. This is a very helpful chapter for anyone working in a field of determinate knowledge that is unavoidably connected to astonishing mysteries. There is, it seems, nothing more practical, knowledge-embedded, and concretely grounded than having a workable metaphysical perspective on the relation of astonishment to the small and intricately beautiful worlds of human knowledge that we construct.

A note in preparation regarding Jeff's chapter. Medicine is a practice intimately concerned with three taboos as regards the "myth of normality" in our increasingly materialist and reflexively triumphant culture; illness, trauma, and death. To speak from within Christian theology, the practice of medicine is a ministry of solidarity with the afflicted, of healing aid, and of consolation in the face of death; it is a Christian practice. Any practice of this nature confronts our mortal vulnerabilities and flux-embedded contingencies, and leads us to the edge of being and to astonishment at the terrifying indeterminacy from where people come and to where they go. Be prepared for a deep and unsettling look at how astonishment and our mortal being interact under the conditions of ministry to the ill, the traumatized, and the dying.

Chapter 9 is by a senior barrister who has a doctorate in legal metaphysics as it relates to the problem of parliamentary hyperactivity. That is, in his doctoral studies Jonathan Horton explored the question of why it is that English and Australian parliaments now make so many new laws all the time. In a similar manner to how Jeff has outlined the implications of William's work for the practice of medicine, Jonathan outlines how an appreciation of the indeterminate high aim of law—justice—relates to the mastery of determinate interpretations of legal precedent. What is striking about this chapter is that none of what Jonathan says is in any sense original, or in any sense unappreciated, but is basic jurisprudence to all the great legal minds of the Western tradition of law and justice, excepting those in our own times. To *us*, however, Jonathan's chapter is profoundly startling. We seem to have forgotten that the making of a judgement is a sacrament, a very this-worldly, tangible, and particular thing that participates in a divine reality that can never be captured by human art or determinate power. Jonathan points out that the habits, rituals, and customs of law in the English tradition still move in accordance with this tacitly sacramental experience of law and judgement, even though the minds of most of our politicians in the creation of new laws, and even many of our lawyers and judges in the interpretation of law, seem entirely unaware of the metaphysical and theological horizon to law. This, Jonathan describes, is the intellectual impact of legal scientism. It is the overcoming of the indeterminate by the determinate. This is a very fine chapter outlining just how concretely and politically vital a respectful metaphysical openness to indeterminable divine gifts is, in our actual lives.

Chapter 10, by Simon Oliver, is an insightful reflection that concerns itself with the manner in which modern science is an inherently technological affair. Looking at how the "final cause" (that is, purpose) of nature itself was re-directed in the seventeenth century from the glory of God to human utility by the founders of modern science, Simon wonders if premodern categories of astonishment are even possible from within a modern scientific perspective. Simon unpacks these historical considerations with the aid of some heavy hitters in this scholarly domain—Hans Jonas, William Eamon, Jessica Riskin—and relates them back to William's chapter, with a focus on that most astonishing aspect of our experience of reality, life. It is a beautiful piece of scholarship that will be of real interest to historians of modern science interested in William's insights into astonishment.

Michael Hanby, in chapter 11, and in the spirit of love and admiration, critiques the way William's chapter seems to allow for the possibility of a meaningful distinction between science and scientism (as is unpacked and defended by Andrew Davison in chapter 5). I must say, when I read Michael's chapter I was delighted, for I thought William will love the metaphysically referenced concept-testing nature of this piece, and William's response will be fascinating. I won't say much more about the argument here, other than that Michael Hanby is one of the most serious English language thinkers writing today whose first concern is the intimate matrix of relationships between metaphysics, theology, and modern science. Michael's argument here follows on neatly from Simon's chapter and these first-order questions about the nature and meaning of modern science are, in my opinion, of great interest to anyone who wants to have a theologically credible understanding of the modern scientific age.

It is worth noting one aspect of the apparent tension between Michael's stance and Andrew Davison's stance. Michael's argument can be read negatively (i.e., as denying the legitimacy of the distinction between science and scientism) or positively (i.e. as arguing that science requires a fuller metaphysical grounding than it presently has—theoretically at least—in order to be more richly and properly scientific). Michael's intentions in making his (negative) critique of the science/scientism distinction are positive (arguing for a better theory and practice of science) so he does not see himself as opposed to science. Indeed, Michael rather sees that the *practice* of science is where the metaphysical richness of creation cannot actually be avoided, even if a *theoretical* outlook that is characterized by reductive empiricism, instrumental rationalism, and a hubris-prone curiosity is unavoidably normative to the way we reflexively approach knowledge itself in our modern, secular, and scientific times. That is, although Michael seems to be arguing in the opposite direction to how Andrew is arguing, whether Michael's concerns actually present a sharp incompatibility with what Andrew is arguing for is not a simple or obvious matter. To explore this demanding matter properly we need a very nuanced and capable thinker. If it is the case that Michael and Andrew can be harmonized in the truth, in a manner that satisfies key features of both arguments, then William is the champion we should call on to see if that aim can be realized.

The twelfth chapter by D. C. Schindler works over the same themes that have been explored by Simon and Michael. The un-nuanced manner in which curiosity has been valorized in modern science, rejecting older cautions and misgivings about curiosity, is a significant component of William's chapter, and this is the theme taken up by D. C. Schindler. The ties between curiosity and the modern scientific interest in knowledge as power require serious attention. For this is not a cosmologically neutral matter. At stake is the question of whether power or love is the first principle of creation. If reality is in fact a function of divine donation, of the gift of Agape, then the full texture of the real cannot actually be known in the instrumental modality of modern curiosity; what the technologist is "knowing" is our own artificial power. So it is Schindler's interest in the metaphysics of the real that motivates his caution towards the mastery epistemology of modern scientific curiosity. If Schindler's concerns are valid, then we will need to rethink what we mean by science itself, if we want to know something at least analogically true about the real. Schindler believes William supports such a radical re-think of modern science, but what does William think about this understanding of his view?

The final chapter in part one of this book—chapter 13—is by the Orthodox priest, theologian, and scholar Isidoros Katsos. Father Isidoros exemplifies, to me, the Orthodox manner in which worship and prayer are the grounds of thinking and engagement with others and the world.

Isidoros brings out the manner in which William's work resonates, in joyous richness, with the Orthodox iconological understanding of the world and our relation to the world. Creation is shot through with the glory of God, and we are riven with astonishment whenever we are touched by a partial awareness of what creation really is. Any real awareness of reality requires a metaphysical appreciation of that which is the grounds of the real, of God. And it is this metaphysical awareness of the unmasterable glory, which is the mark and presence of God in, beneath, and beyond all creation, that saves us from the spiritual temptations of reductively determinate knowledge. To both Isidoros and William, science has its rightful place only when its determinate categories are held as non-ultimate, and subordinated to worship. This right appreciation of determinate knowledge is lost without a properly doxological humility and joy. Only ortho-doxology can save science from being a spiritual trap, a fall into the relentless hubris of a restless ontological unreality. The finite and the temporal rests between the infinity of Nothing and the beyond-infinity of God's eternal Being, and our minds will always be restless until they rest in the *metaxu* (the between) of where God has placed us and astonishingly meets us. Accepting this doxological rest is what it means to embrace being a creature. Curiosity is only spiritually safe—and joyfully limited—to the worshiping mind. This is a profound and astonishing chapter.

◆ ◆ ◆

Part Three of this book is William's response to the engagements that our twelve interlocutors provide. This is a treat. Weaving the fabrics of all responses together, we see further into aspects of each response, and gain further illumination on the original chapter. Via this passage new light is shed on the mystery of astonishment, and in that light we start to dimly see what better relations between the different modes of knowledge and wisdom could look like. This final chapter is incredibly helpful as we wrestle to think about how to best respond to the challenge of scientism and the value of good natural philosophy. This chapter helps us move towards a richer way of trying to think about and practice natural knowledge and the light of revelation and wisdom in ways that are more adequate to the mystery and wonder of the real.

Bibliography

Desmond, William. *The Voiding of Being*. Washington, DC: Catholic University of America Press, 2020.

Harrison, Peter. *The Territories of Science and Religion*. Chicago: Chicago University Press, 2015.

PART ONE

1

The Dearth of Astonishment

ON CURIOSITY, SCIENTISM, AND THINKING AS NEGATIVITY

William Desmond

Curiosity and Scientism

WHAT LOOKS LIKE AN insatiable curiosity seems to drive the scientific enterprise. In principle no question seems barred to that curiosity. There seems something limitless to it. Are there recessed equivocities in this drive of limitless curiosity? How does one connect this question with the issue of scientism? Do these equivocities have anything to do with the possibilities of misunderstanding the limitlessness of our desire to know? Are there temptations in curiosity itself to misform the potential infinitude of that desire to know? Is it possible for certain configurations of curiosity to produce counterfeit doubles of wonder? Has the scientistic impulse something to do with the temptation to this malformation? Has this something to do with how the infinity of our desire to know turns away from the true infinite—turns itself into the truth of the infinite?

Scientism bears on the place of science in the life of the whole from which there is no way to abstract science entirely. In the main, scientism understands science to be capable, in principle, if not in practice, of answering all the essential perplexities about the world and the human condition. I want to reflect on the temptation to contract the meaning of wonder in the scientistic interpretation of scientific curiosity. I will ask if this contraction of wonder paradoxically can be tempted with a limitless self-expansion out of which a kind of idolatrous knowing can come. I want to explore a connection with the determinateness or determinability of being, a connection present also in the idea of thinking as negativity. I want to explore a dearth of ontological astonishment in all of this.

We need to distinguish different modalities of wonder: wonder in the modalities of astonishment, perplexity, and curiosity. Curiosity tends to be oriented to what is determinate or determinable, perplexity to something more indeterminate, astonishment to what is overdeterminate, as exceeding univocal determination. The determinate and determinable curiosity of science, while capable of showing a face of unlimitedness, is not to be equated with the fullness of wonder. Throughout my reflection, the intimate relation of determination and self-determination will be of importance: there is a configuration of wonder overtly tilted to determination but which, more intimately, is ingredient in a project of self-determination that wants to bring itself to full completion. I am interested in asking if our curiosity, under scientistic influences, can come to communicate idolatrous outcomes. There is a religious issue here of how, in the name of high ideals, we secrete counterfeit doubles of God. The open infinitude of our seeking might be taken to reflect our *capax dei*,[1] but there is a difference between a capacity for the divine, a capacity that finds in itself something divine, and a capacity that wills itself to be (the) divine. I am interested in asking if the scientistic contraction of curiosity creates a counterfeit infinity.

I offer some general observations about scientism. First and importantly, scientism is not science but an interpretation of science. The questions here are inevitably of a philosophical nature, with repercussions for the full economy of human life. In that full economy, it is unavoidable that we ask about the place of science in the life of the whole. There is no way to abstract science entirely from this sense of the whole, even though this sense may be more or less recessed in our self-consciousness.

Scientism is a philosophical interpretation of the whole, though it takes place within the whole. How is this life of the whole at stake in this interpretation? The answer to the question is that what we call science, at least in principle, has the resources to supply all answers to all the meaningful questions. It has not only the resources, but it has also the privilege and right to make this claim. There is something privileged about science on this interpretation. One of the considerations relevant here is a strong tendency

1. Meaning: "capacity for God."

to univocalize the meaning of the scientific impulse in terms of a master science that will offer this master key, or the hope of it, as opening all doors of remaining mystery, as overcoming all residual ignorances.

Of course, there are different scientisms, depending on the state of the current science, and depending on which particular science seems to be in the ascendency in the epistemic stakes. One thinks of older scientisms in the eighteenth century with some of the *philosophes*; one thinks of how in the nineteenth century a variety of materialisms vied for this ascendency.[2] One might associate determinisms of various sorts with the temptation to seek the one true cause, or set of causes, that will allow the scientific determinability of all being. One thinks of Freudianism: a certain understanding of eros as the one source of explanations, and if not explanation, illumination.[3] In the twentieth century, there have been various "master sciences." We now forget how Marxism was once considered so. One forgets how linguistics came on the scene as offering the power of the keys. The *potestas clavium*[4] has passed variously to different sciences and more lately, perhaps, to genetics, for its reign as a brief god. Genetics offered a second coming to Darwinism.

How often do we hear it told that everything reveals adaptive evolutionary significance? Evolutionary psychology: evolution "wires" us so; neuropsychology morphing into neurophysiology will illuminate all. Cybernetic scientism: the mind is to be likened to a computer, and of course, it is like it, since the computer is the product of mind. I think there is a droll lesson from evolutionary aesthetics—it seems we are adapted to aesthetic perceptions laid down deeply in us from our time when we were getting by on the African savanna. It seems that the aesthetic scene most loved by humans is a variation of a scene on the savanna, with preference for some trees and some animals in the distance, a variation on a pastoral scene of some sort. The postmodern avant-garde is not amused to think that the earliest human beings have something in common with the "booboisie," H. L. Mencken's contemptuous phrase (dating from early in the twentieth century) for ordinary middle-class Americans. Despite the tastes of the advanced aesthetes, the aesthetic psychology that evolutionary unfolding

2. One thinks of Darwinism; also of Marxism as a dialectical scientism; even of philosophic idealism as a kind of (speculative) scientism, especially in the hands of a Hegel. In the latter, as we shall see, there are complex qualifications, as his sense of *Wissenschaft* distinguishes philosophy from one of the particular sciences, which to Hegel are finite sciences. Still determination serves self-determination, a project marking most scientisms.

3. See the salutary and thought-provoking Crews, *Freud: The Making of an Illusion*.

4. Meaning: "Power of the keys."

seems to trump for is a deplorable love of kitsch. One thinks too of the transhumanism of a kind of cybernetic Nietzscheanism as having scientistic seeds.[5] I recall an advocate of transplanting minds into durable matter. The soft and vulnerable flesh has no future. Transplanted into steel, we could be shot out into space as, say, chairs.[6]

How we understand the culture of knowing is in question. Culture: the deepest and widest sense in which the cultivation of the desire to know is nurtured; of course, culture and *cultus* are related members in the same family. *Scientia*: there is a more inclusive sense of knowing that in some ways is coincident with human being as a creature capable of coming to mindfulness of itself and being. This can include both self-knowing and knowing what is other. Obviously, much then hangs on what it means to know. There is also the relation between this more inclusive sense of knowing and the plurality of specific and determinate forms of knowing, among which are the knowings of the particular sciences.

With reference to this more inclusive sense, diverse forms of knowing are seeded in an original wonder. Indeed, there is something about this wonder that is prior to the desire to know, insofar as such a desire is already a specification of an original porosity of being in us, an original porosity that is open to being. The desire to know takes shape in this original porosity, and the shapes it takes can be plural. Without the more original porosity to being in our porosity of being there is no desire to know, and no more specific determination of that desire. Curiosity about being and the determinate beings and processes to which we are opened is subsequent to, secondary to, this more original porosity. Scientific curiosity, such as we have come to know it in a long history of its diverse unfoldings, is derivative from this original porosity. But the diverse unfoldings can forget the maternal porosity out of which they are born, and take on an energy of determination that rides over the original givenness of this porosity.

Curiosity is born from original wonder but it is tempted to think of itself as self-born, self-activating. Its drive for determination is tempted to think of itself as self-determining. It is hence also tempted to forget the original heteronomy out of which it is birthed. Self-determination and the

5. Do not forget that many early Bolsheviks were as much enamoured of the *Über-mensch* as were the National Socialists who were attracted to Nietzsche. See the science fiction, better scientistic fiction of Harari, *Homo Deus*, the final chapter of which ends with the prophesy of the best coming religion: "The Data Religion," the coming all-conquering religion of the future, given that intelligence is now said to be no longer dependent on consciousness, and presuming that we have the right algorithms and "data" enough to feed into our cybernetic machines.

6. I prefer flesh, albeit vulnerable and frail.

determinability of being then tend to go hand in hand. The desire to know is driven by the desire to make given being more and more determinate; and this is tied up with our own desire to be the ones who determine the intelligibility of being, perhaps sometimes for its own sake but mostly for purposes of serving our own self-determination. The projection of curiosity to determine being as univocally as possible becomes the project of our self-determination as more and more absolutely autonomous. Our self-determination defines itself as more and more for itself, the more it can stamp its impress on given being as determined by it.

Parenthetically: Heidegger came in for a good deal of criticism when he said (in *What Is Called Thinking?*) that "science does not think [*Die Wissenschaft denkt niet*]." In an interview explaining his view, Heidegger stressed the lack of self-reflection involved in science.[7] Surely he is not wrong about that, and in agreement with Husserl: science does not make itself its own object of reflection. But surely also there is more to the issue. The lack of thinking is not remedied by the supplement of self-reflection. The mystery of being in its given otherness is also at issue. Heidegger was trying perhaps to remedy the deficiency when he rightly advocates a dialogue of thinking and poetry. Even then was he perhaps *too dyadic* in his thinking? When Heidegger speaks of truth as *alētheia* and *orthotes* he is making an important point, an indispensable point in some ways. Often philosophers have focused on the second in terms of a correspondence of propositions with an already given reality, to the negation of a prior sense of "being true" which is presupposed by propositional truth. This prior sense is not a matter of propositions only. Heidegger is opening up the meaning of the nonpropositional as more original, and as at the origin of the propositional. I approach this issue somewhat differently in terms of the porosity of being. We need to avoid univocal unity as absolute, but also dyadic contrast as ultimate, as well indeed as the dialectical deconstruction of univocal unity and dyadic equivocity. The original porosity is better named and articulated in terms of a metaxology of being.

By comparison with the univocal, dyadic, and dialectical approaches, we need to think of the distinctions of system and being systematic, and of their connections with the trans-systematic. There is a metaxology of art, religion, and philosophy, beyond dyadic thinking. The same applies to the significance of science. Husserl said that he aimed to destroy the kind of thinking that he later, too late, sees Heidegger as doing. He wanted to make that impossible. Against naturalistic scientism Husserl wants to open

---

7. YouTube. "Wissenschaft denkt nicht," accessed May 16, 2019; available at https://www.youtube.com/watch?v=HwuSmN5ptGA. Heidegger, *What is Called Thinking?*, 8.

up a transcendental philosophy whose path in and through "self-reflection" opens up also to being. Is it too transcendental? Is it possible to end up in a kind of transcendental scientism? I would pose the question in relation to the first crystallization of the impulse to transcendental absoluteness, namely the passage of philosophy from Kant to Hegel. Hegel and his system: is there not a danger of a kind of speculative scientism? The anti-idealists like Schopenhauer and Nietzsche do not offer consolation, in the end: Schopenhauerian monism of will; Nietzschean monism of will to power. It is the univocalization in their monism that is the cousin of the scientism.

The senses of being are at stake. In various forms of scientism the underlying presupposition is that being is determinate, and in a manner that invites science to make it as univocally precise as possible. Again the ideal of science as being as univocal as possible, bearing on the meanings of truth and intelligibility, is not separable from the project of self-determination. We would make being determinate in a scientistic spirit in order to further the project of our self-determination, via the determination of being as other to us. Cybernetic transhumanism strikes me as one of the latest forms of this.

I believe we have to take note of three modalities of wonder, keeping in mind that wonder is not a univocal concept. It is not first a concept at all, but a happening, and as a happening it is plurivocal. The three modalities are internally related to each other, but they reveal a different stress in the unfolding of our porosity to being. If we do not properly attend to these different stresses, we can mistakenly think that all wonder is subsumable into the curiosity that makes of all being an object of determinate cognition. This subsumption might consume curiosity, but it is the death of wonder. This is the temptation of the scientistic impulse, a temptation that in being enacted is justified in terms of a particular teleology of the desire to know. To the contrary, wonder is not to be solely reconfigured as voracious curiosity that spends itself in ceaseless accumulation of determinate cognition(s).

Between Indeterminacy and Overdeterminacy

To articulate the matter I find it helpful to invoke the notions of the *indeterminate*, the *determinate*, the *self-determining*, and the *overdeterminate*. They are relevant to understanding the dearth of ontological astonishment and the contraction of curiosity in scientism. They have a bearing on how we orient ourselves in thinking of being as given. I will say something about these notions, and then I will turn to different modalities of wonder, before returning again to the contraction of curiosity in scientism.

Looking back to ancient metaphysics, the indeterminate tended to be seen in the light of the absence of intelligibility. By contrast, determinacy or determinability, so to say, keeps at bay the formlessness of the indeterminate (*to apeiron*). The epistemic process by which being is made intelligible involves the movement of thought from the indeterminate to determinacy, the former being left behind in the process. We begin with an indeterminate wondering, pass through a more definite questioning or inquiry, and end with a more or less determinate answer to a well-defined question. In truth, the unfolding here is more complex than a teleological process from indeterminacy to determination. We need to invoke the self-determining and more importantly the overdeterminate, in a sense to be addressed. The point is not to negate the indeterminate but to offer the more affirmative sense suggested by it: no longer as a formlessness to be overcome but a "too muchness" in being that calls forth our ontological astonishment.

In everyday realism we think that things and processes have a more or less fixed and univocal character, and that this constitutes their determinacy. Nevertheless, determinacy cannot be understood purely in itself; it refers us to the outcome of the process of determination, a process not itself just another determinate thing. We tend to separate the determinate outcome from the determining process, and so take what is there as composed of a collection of determinate things. Determinacy is bound up with the fact that things and processes do manifest themselves with an immanent articulation and relative stability. Whether that immanent articulation can be expressed entirely in univocal terms is an important question. If we put the stress only on univocity, we can cover over the process by which the determinate comes to be determinate. Equivocal, dialectical, and metaxological considerations enter into a fuller account of determinacy.

When indeterminacy is invoked it is often by contrast with determinacy. This might seem to be essentially a privative notion, referring us to the absence of determinate characteristics, and thus hard to distinguish from what is void. I suggest a more *affirmative* understanding, one referring us to the matrix out of which determinate beings become determinate. As a kind of predeterminate matrix, this would reveal determining power in enabling the determinate things that come to be. This more affirmative sense makes us think of the idea of *overdeterminacy*. Void indeterminacy refers us to an indefiniteness that is only the absence of determination, rather than the more fertile matrix out of which determinacy can come to be. These two senses of the indeterminate are often mixed up. If overdeterminacy is presupposed by indeterminacy, our general tendency to oscillate between the indeterminate and the determinate will be seen not to go far enough. If determinacy is often correlated with univocity, and indeterminacy with equivocity, we need further dialectical and metaxological resources to do full justice to what is at play.

An additional consideration concerns the notion of *self-determinacy*. This refers us to a process of determination in which *the unfolding recurs to itself* and hence enters into self-relation in the very unfolding itself. This is particularly evident in the case of the human being as self-determining. The notion cannot be fully understood without reference to the ideas of the indeterminate and the determinate. Frequently self-determination is seen as the determination of the indeterminate in which a process of selving comes to achieve a *relationship to itself*. Again the human being is the most striking example of this, and particularly in modernity the idea of self-determination has received central attention. But both self-determinacy and determinacy refer back to something that cannot be described in the terms of self-determination or determination. This something other is not just the indeterminate understood in the privative sense, but the more affirmative sense that is the overdeterminate. Self-determinacy comes to be out of sources that are not just self-determining. Our powers of self-determining are endowed powers. There is a receiving of self before there is an acting of self. This makes the process of selving porous to sources of otherness that exceed selving.

This matter can be illuminated in a number of ways, but I want to refer to the speculative logic of Hegel where the triad—the indeterminate, the determinate, and the self-determining (sometimes via reciprocal determination)—governs this entire process as ultimately mediating itself in a self-becoming in which the other to the selving is the (self-)othering of the selving itself. The overdeterminacy does not enter systematically into the articulation of this understanding. I want to suggest that the overdeterminacy, as the affirmative sense of the indeterminate and not as the negative sense of the indefinite, refers us to the enabling matrix that makes possible determinacy and self-determination. There is something prior to the determinate but not a mere indefiniteness. It has an excess more than all determinations, as well as more than what we can subject to self-determination. There is a "too-muchness" that has a primordial givenness that enables determinacy, that companions our self-determination, and yet also exceeds or outlives these. It is not to be equated with overdetermination understood as necessitation by an excess of determining causes. It allows the possibility of the open space of the indeterminate, and hence is not hyperbolic determinism, but hyperbolic to determinism in enabling the endowment of freedom. If Hegel's dialectic tends to be defined by the triad of the indeterminate, the determinate, and the self-determining, metaxology exceeds this triad in the direction of remaining true to the inexhaustible overdeterminacy. This inexhaustible overdeterminacy is multiply incarnated, for instance, in great artworks, persons, or communities.

Ontological Astonishment and Hegelian Negativity

Let me now say something about Hegel's notion of thinking as self-relating negativity, because by this means we can see the connection between determination *qua* process and self-determination, as governing the whole purpose of the move from the indeterminate to the determinate. If to be intelligible is to be determinate—indeed, *to be*, properly speaking, is to be determinate—the status of determinacy and determination opens itself for question. How comes the determinate to be determinate? How does it come about that being as determinate is determinable by thought and hence rendered intelligible? The issue of a *becoming* determinate is at stake, not just some entirely static sense of being. I see Hegel's connection of thinking with determinate negation, or more generally with subjectivity as self-relating negativity, as answering such questions. What is simply given to be is not intelligible as such; it is a mere immediacy until rendered intelligible, either through its own becoming intelligible, or through being made intelligible by thinking. Thinking as negativity moves us from the simple givenness of the "to be" to the more determinately intelligible; but the former (the "to be") is no more than an indeterminacy, and hence deficient in true intelligibility; until this further development of determination has been made by thinking as negativity. A further complication in Hegel's view is that thinking as process of negation is not only a determining; it is in process toward knowing itself as a process of self-determining. Hence his more complex description: self-relating negativity. The operation of negation is not only a determination of what is other to the thinking, it is the coming to itself of the thinking process. In that sense, the return of thinking to itself, in the process of determining what is other, is not just making determinate, it is *self-determining*. The determining power of thinking in negativity is hence inseparable from Hegel's understanding of the meaning of freedom. But there is an overall logic that governs the movement of thinking as negativity: thinking moves from indeterminacy to determination to self-determination.

In all of this there is a dearth of ontological astonishment, I claim. For instance, given being as a mere indeterminate immediacy can barely be said to be, and even less said to be intelligible, until rendered so by determining thinking, which mediates by negativity. Hence, being becomes the most indigent of the categories that is all but nothing, until thought understands that it has already passed over into becoming. I do not want to rehearse the famous opening of Hegel's *Logic*, but want to suggest, among other things, that Hegelian negativity, via a logic of self-determining thought, is born of and leads to a dearth of ontological astonishment. Instead of a sense of being as the marvel of the "too much," we find rather an indigence of "all but nothing." To further illuminate the issue, we need to distinguish between different modalities of wonder: first a more primal ontological astonishment that seeds metaphysical mindfulness; second a restless perplexity in which thinking seeks to transcend initial indeterminacy toward more and more determinate outcomes; third, more determinate curiosity in which the initiating openness of wonder is dispelled in a determinate solution to a determinate problem.

Determining thought answers to a powerful curiosity that renders intelligible the given, rather than to a primal astonishment before the marvel of the "to be" as given—given with a fullness impossible to describe in the language of negativity, though indeed in a certain sense it is no thing. Heidegger, for instance, has a truer sense of this other nothing. My focus is less defending Heidegger as to suggest the need to grant something more than a *logos* of becoming and self-becoming—there is an event of "coming to be" that asks of us a different *logos*. It asks of us a different sense of being, a different sense of nothing—not the nothing defining a determinate process of becoming, or a determining nothing defining a self-becoming. Rather, a nothing in relation to which a coming to be arises—a coming to be that is more primal than becoming. In a way, we can say nothing univocally direct about this nothing; rather we need to attend to how becoming and self-becoming presuppose this other sense of coming to be. A sense of this is communicated in the happening of a primal astonishment before the happening of the "to be" as overdeterminate. In light of it every process of determination and self-determination is secretly accompanied by what it cannot entirely accommodate on its own terms. This granting of the overdeterminacy of the "to be" has significance in relation to the dearth of ontological astonishment in thinking as determinate negation, or self-relating negativity. It has a very important implication for the practise(s) of metaphysical thinking that try to stay true to metaphysical wonder in the mode of primal astonishment.

Being Overdeterminate: Wonder as Astonishment

I now want to both deepen and widen our investigation by saying something about the three modalities of wonder.[8] We have to keep in mind that these

---

8. In this and following two sections, I am drawing from, and adapting, a longer discussion in my *The Intimate Strangeness of Being*, ch. 10.

three are internally related to each other as they reveal a different stress in the unfolding formation of our porosity to being. They are not three epistemic layers stacked one on top of the other. Taking form in the porosity, giving form to the porosity, there is a fluidity of passage of one into the other, indeed of permeability and mingling of one and the other. As Joyce says of the waters of Anna Livia Plurabelle, there is a "hitherandthithering" flow to them. But there are different stresses in the fluctuation, and if we do not properly attend to them, we can mistakenly think all wonder is subsumable into the curiosity that makes of all being an object of determinate cognition. This subsumption might consume curiosity, but it is the death of wonder. Wonder is not to be solely reconfigured as voracious curiosity that spends itself in ceaseless accumulation of determinate cognition, even unto the extreme of the counterfeit infinity of scientism. Equally, there is something other to thinking as self-relating negativity. There is something about wonder that exceeds self-relation. We do not possess a capacity for wonder; rather we are capacitated by wonder. This capacitation is not determined through ourselves alone. We alone cannot just will to bring it to be, or will to bring it to life again, if its power dies down. We do not just exercise self-determination over it. There is a given porosity of being that endows us with the promise of mindfulness. The self-determination of our powers of minding are derivative of this endowment.

Turning to the first modality of wonder, it is impossible to describe this astonishment in the language of negativity. There is a wonder preceding determinate and self-determining cognition. Wonder before the being there of being and beings is precipitated in this astonishment. This has not to do with a process of becoming this or that but with porosity to the "that it is at all" of being. That being is, that beings have come to be at all—this is prior to their becoming this or that, prior to their self-becoming. In a certain sense, all human mindfulness is seeded in this astonishment.

We have to be cautious here, as "wonder" is often seen in a way that is too subjectivized. It is the feeling of "gosh" or "wow" that is said to be experienced before what surprises us. One of the reasons I use "astonishment" is because it captures better the ontological bite of otherness. There is *the stress of the emphatic beyond expectation*. We say "The wonder of it is . . ." and we mean to take note of a happening beyond expectation, one in which the surprising has been communicated with this stress of the emphatic. There is something of the blow of unpremeditated otherness in being struck by astonishment. The otherness seems to stun us, bewilder us, even stupefy us. We seem deprived of self-possession when we are *stricken* with a kind of amazement. We seem to be overcome with a kind of ontological stupor.

All of this seems to be rife with a kind of negation—not our negation but our being negated. It is not thinking as negation. Something of the affirmative "too-muchness," the overdeterminacy of being has been communicated in the astonishment. This is not free from a dimension of ontological frailty in that the surprising event might not have come to be, it might not have been at all. And yet it has eventuated, despite the possibility that it might not have been at all. We awake on a boundary between being at all and possibly not being. Our porosity to the overdeterminacy of the happening is linked with the intimation that the "too-muchness" is also a kind of "no-thing." The first appropriate way to think of this is not in terms of negativity. This might come later. First, in the opening of porosity, something strikes into us, while at the same time taking us beyond ourselves. Out of this, but derivatively, the self-transcending of thinking is possibilized. We see the inappropriateness of speaking of astonishment as just a subjective feeling. We are moved into the space of a *metaxu*[9] where we are enabled to go toward the things, because in the porous between there is no fixation of the difference of minding and things. And so it is not quite that we go from our minds to the things, but more so that the things come to mind. The things come to mind. This is what "beholding from" also entails.[10]

Instead of thinking as negativity, already even before we more reflectively come to ourselves, there is the more primal opening in astonishment. There is a prior porosity in which there is no fixed boundary between there and here, between outside and inside; there is passage from what is, passage into an awakening of minding, which is opened to what communicates itself to us, before our own efforts at self-determining cognition. We do not open ourselves; being opened, we are as an opening. Astonishment awakens the porosity of mindfulness to being, in the communication of being to mindfulness, before mind comes to itself in more determinate form(s). It is for this reason also that I would correlate it with a more original "coming to be"

9. Meaning: in-between.

10. My use of the phrase "beholding from" is derived from William Wordsworth's "Lines Written above Tintern Abbey," i.e., "Therefore am I still / A lover of the meadows and the woods, / And mountains; and of all that we behold, / From this green earth; of all the mighty world / Of eye and ear." Wordsworth, *William Wordsworth*, 134.102–6. The phrase occurs after Wordsworth speaks of the spirit "that rolls through all things" and before the lines that mention "of all the mighty world / Of eye, and ear,—both what they half create, / And what perceive." Was Wordsworth here caught in the tension of a kind of idealism, not fully true to the "from"? The problem is evident in Coleridge's lines in "Dejection: An Ode": "O Lady! we receive but what we give, / And in our life alone does Nature live: / Ours is her wedding garment, ours her shroud!" Coleridge, *The Major Works*, 114. If we receive only what we give, there is no "beholding from." If there is "beholding from," we receive before we give.

prior to the formation of different processes of determinate becoming and the arrival of relatively settled beings and processes.

Because all determinate thinking already presupposes it as having happened, we find it difficult to think this more original porosity of astonishment. It is not yet determinate knowing, yet all determinate knowing proceeds from it. Yet it can be communicated. I behold the majestic tree and exclaim: "This is astonishing!" I am not projecting my feeling; rather, the tree is coming to wakefulness in me, while I am being awakened by the tree, and I am awakening to myself, in a more primal porosity. The striking otherness of this blossoming thereness has found its way into the intimacy of my receiving attendance. This astonishment is a porosity prior to intentionality; it is not a vector of intentionality going from subject to object; it refers us back to a patience of being more primal than any cognitive endeavor to be. Porosity might seem like negativity in that it cannot be reduced to this or that determination, and allows dynamism and passage. At issue is not thinking as negation but rather a mindful *passio essendi*[11] prior to and presupposed by every *conatus essendi*[12] of the mind desiring to understand this or that. It is not first that we desire to understand. Rather, we are awoken or become awake in a not-yet-determinate minding that is not full with itself but filled with an openness to what is beyond itself. Is it permissible to say one is filled with openness, given that such a porosity looks like nothing determinate and hence seems almost nothing, even entirely empty? Being filled with openness and yet being empty: this is what makes possible all our determinate relations to determinate beings and processes, whether these relations be knowing ones or unknowing. Thinking understood primarily as negativity does not have enough of this porous patience, even though its endeavor to know ultimately derives from it.

One might object that the desire to know is a drive to determination, a drive that when it comes to know itself becomes also more self-determining. It is a well-rehearsed theme that philosophy begins in wonder and Aristotle is often cited: "All men desire to know" (*Metaphysics* 982b11). Aristotle sees the connection of marveling and astonishment when he reminds us of the affiliation of myth and metaphysics, and also the delight in the senses. Nevertheless, the desire to know is understood essentially as a *drive to determinate intelligibility*, which on being attained dissolves the initial wonder setting the mind in motion. The end of Aristotle's wonder is a determinate *logos* of a determinate somewhat, a *tode ti*. This end is the dissolution of wonder, not its deepening. I find it significant that Aristotle calls on *geometry* to

---

11. Meaning: the passion of being.
12. Meaning: the endeavor of being.

illustrate the teleological thrust of the desire to know (*Metaphysics* 983a13). Geometry can be taken as exemplary of determinate cognition: when the problem is solved, the wonder is extinguished and surpassed.

Plato is more suggestive in the *Theaetetus* (155d3–4) when *thaumazein* is said to be the *pathos* of the philosopher. *Pathos*: there is a patience, a primal receptivity. This is at the other extreme to the self-activating knowing we expect from Kant and his successors, both idealist and constructivist. There is a pathos more primal than activity, a patience before any self-activity. There is no going beyond ourselves without the more primal patience and receiving. Patience is often denigrated as a servile passivity supposedly beneath our high dignity as self-activating powers. And yet no one can self-activate into wonder. It comes or it does not come. We are *struck* into wonder. "Being struck" is beyond our self-determination. We cannot "project" ourselves into "being struck." It comes to us from beyond ourselves.

While I will return to curiosity, let me offer a brief comparison between it and primal astonishment. I would correlate curiosity with a determinate cognition of a determinate somewhat (*tode ti*) or "object." By contrast, in astonishment an "object" as other does not simply seize us and make us "merely" passive. What is received cannot be thus objectified. What seizes us is the offer of being beyond all objectification, and the call of truthfulness to being. This is not first either subjective or objective, but trans-subjective and trans-objective. "Trans": we witness a *crossing between* "subject" and "object" and an intermedium of their interplay that is more primordial than any determinable intermediation between the two. The happening of this "being-between" reveals a porosity beyond subjectification and objectification and we are beholden to what eventuates in this between, making us answerable to its truth in our own being truthful. In the intimate strangeness of the porosity an excess of being flows, and overflows toward one.

We could say there is something *childlike* about this, but this is not to say it is *childish*. Childlike we find ourselves already in the porosity of being: astonishment is not produced, it *opens us* in the first instance, and there is joy in the light. The child *lives* this primal and elemental opening; hence wonder is often noted as more characteristic of earlier stages of life, often accompanied by an asking of the "big questions." Subsequent developments of curiosity and sophisticated scientific knowing are seeded in the primal porosity but what its grant enables we too quickly take for granted. The maternal porosity can be long forgotten when the project of science comes more fully on the scene. When the child points to the night sky and murmurs—"Look, the moon!"—the astonishing has won its way into its heart. Later, the astonished child is recessed, even driven underground, in the curious project of (say) space exploration which lifts off the earth.[13]

The more primal porosity is at the origin of all modalities of mind, but as intimate with the giving of the first opening, it can be passed over, covered over. Because it enables the passage of mindfulness it can be passed over, for we come to ourselves more determinately in this passing. It is first a happening and only subsequently is what happens gathered to itself in an express self-relation. In the latter, we risk the contraction of what the first opening communicates. We can contract the opening of the porosity to just what *we* will grant as given. The point is relevant to wonder in the modalities of perplexity and curiosity, and in modes of minding that are determinate and self-determining. The porosity is prior to determinate and self-determining cognition, and is neither of self or other but happens as the between space in which, and out of which, come to be a variety of determinate and self-determining mindings of being. The latter are derived, not original. What is more original is the between of porosity.

Being Indeterminate: Wonder as Perplexity

I think of perplexity as a second modality of wondering whereby we pass from the overdeterminacy to a mingling of the indeterminacy and the determinate. If one were to refer to thinking as negativity, it would be more appropriate to consider it in connection with perplexity. When we are perplexed we are often beset by doubt and uncertainty. Something puzzles us and we cannot quite solve the puzzle. We see the word *"plexus"* in perplexity, and what this suggests is a plaiting, a twining, an entanglement. The word "com-plex" points to something intricate and difficult to unravel: a knot we struggle perhaps to untangle. We find ourselves plagued by perplexity, tormented with some vexing matter. We are not sure what to think, and there is no serenity of mind. More often there is the threat of something bewildering, troubling, perturbing.

How does one think of perplexity as arising out of first astonishment? In astonishment we come to be granted as coming to mindfulness both of the overdeterminacy of other-being and a kind of indeterminacy opening in the porosity of our own being. We are granted as being for ourselves

---

13. My thanks to Catherine Pickstock who responded to some of the remarks in this chapter in a presentation given in the series "Grammars of Wonder," organized by Philip McCosker at the von Hügel Institute, St. Edmund's College, University of Cambridge, February 23, 2018. I particularly note her acute remarks on Sylvia Plath's "The Moon and the Yew Tree."

now in more express contrast to the overdeterminacy in which we more originally participate. We risk being overwhelmed by that overdeterminacy and want to reduce it to a more determinate measure, allowing us also to take its measure, to whatever extent is possible. Perhaps here some more express sense of thinking as negation can come into the open. The "too-muchness" of given being can seem to *oppress* us.[14] The intimation intrudes that we cannot be its full measure; yet we want to know it in full measure. A troubled disjunction arises: we do not know, we would know, we know we do not know. We endure the stress of a baffling difference between what we know is too much for us and our intimately known desire to know just that "too-muchness." In the baffling difference we are torn between our desire to know and our intimate knowing that the perplexing is too much for us. We are tempted to diminish the stress in seeking a knowing that reduces the "too-muchness" to proportions that allow us to appropriate its difference. The desire to know can be developed as our way of subjecting the given "too-muchness" to our measure, that is, to the *proportionate* measure of ourselves as knowers. That said, wonder as perplexity is recurrently haunted by faces of otherness that are just so as *disproportionate* to the determinate measure of our determinative cognition.

Perplexity might be correlated with the *equivocity of being*. We are enthralled by the play of light and darkness, the chiaroscuro of things and ourselves. Enthralled as both enchanted by and in thrall to the dark light of unformed things and things forming, of ourselves formless and seeking form, of all things enigmatic and intimating, of ourselves as the most baffling of beings. This is not the reverse of astonishment but our awakening to the sometimes perturbing ambiguity of the overdeterminacy. Perplexity affects both the sides of self-being and other-being. Other-beings and selvings come from formlessness beyond form, are themselves as forming and coming to form, and finally point beyond themselves and all finite form. The equivocity can invade us with foreboding in the face of the mystery of life. Filled with dismay, we can be driven to distraction, can even be driven mad. At the same time, this perplexity can awaken an urgent *seeking* for what is true in all significant art, in all intellectually honest philosophy, in all spiritually serious religion. Mostly, however, the seeking has no fancy

14. Edmund Burke sees a connection of astonishment with horror: "astonishment is that state of the soul, in which all its motions are suspended, with some degree of horror." *A Philosophical Enquiry into the Origin of Our Ideas of the Sublime and the Beautiful*, part II, section 1. He is not wrong but he is not entirely right either. In terms of the analysis here, the sense of horror becomes more overt *on the turn* of wonder from first astonishment to perplexity.

names, as ordinary persons in accustomed community, mostly out of the limelight, seek to tread the way of truth.

There is a saturated equivocity to perplexity: the *doubleness* of being, *both* the dismaying destitution of not-knowing *and* the ignorance of a voracious desire to know. In perplexity we can wake up to ourselves before and beyond the determinate desire to know that we most often find in curiosity. Perplexity has something more primitive than what we normally call the desire to know, as well as something potentially touching transcendence as other. This more primitive perplexity is gestated in the intimate selving that comes to be out of the original ontological porosity. It is trans-objective and trans-subjective, and hence more than selving alone with itself. As being awakening out of the porosity, it is already an equivocal way of "being with" what is other than selving.

One could say in respect of the equivocal play of light and darkness, we are in a condition reminiscent of Plato's Cave. In perplexity, however, we are not in the Cave as prisoners who do not know they are prisoners. These latter do not know perplexity as an awakening. The perplexed have an intimation of being held in check by something too much for their own power. Being perplexed we can be nonplussed by the saturated equivocity of being. In moments of more porous mindfulness lucidity can break into perplexity, and we know that there is light. The light might be twinned with darkness, but that does not make it any less the light. *We ourselves* are double: perplexed between the burden of the mystery and the godsend of light that gives ontological uplift.

Perplexity can be haunted by *horror*: ontological horror before the being-there of being in its excess to our rational measure. It is not often enough remarked that in the Cave we can turn *downwards* as well as upwards. There can be a debilitating perplexity where we have the feeling of *being blocked from ascending into the light*. We seek to find the light, but instead we find ourselves darkened. We find ourselves in the dark in the very seeking for light itself. We lack the night vision of the wise. We are not the measure of the light; we are also not the measure of the darkness. We find we cannot go up on our own. In perplexity we find ourselves falling, though we do not want to fall.

I find it interesting that in the same dialogue, the *Theaetetus*, where Plato singles out *thaumazein*, the aporetic is also stressed again and again. The experience of *aporia* or impasse can show the working of extreme perplexity in thought. *Aporia*: a lack of *poros* and we cannot find a way across a gap. The aporia does not preclude further and new thinking, but it can also be addressed by myth or likely stories. Univocal theories are not enough. I am also put in mind of Kant: there are metaphysical questions we cannot avoid raising but cannot also answer or put to rest in a univocal science. There is the zigzag method for perplexity (Aristotle and Kant?): "on the one hand, this," "on the other hand, that." Perplexity calls to mind a fever where we restlessly rock this way, rock that way, unable to find peace. Thinkers like Kierkegaard, Nietzsche, and Dostoevsky had experience of this fever, sometimes suffering the unavoidability of thinking as a kind of sickness. They feel cursed by reflection as an unease, as a disease.

What here of thinking as self-relating negativity? Such thinking is inclined to claim that through its own self-determination it can counter-act the falling into equivocity by means of its progressive determination of intelligibility. The Hegelian way of doubt (*der Weg des Zweifels*—notice the reference to the double, *zwei*) will overcome radical equivocity through its own self-accomplishing skepticism.[15] In accomplishing itself, skepticism overcomes skepticism, gives up its vagrancy (Kant described the skeptic as a nomad), and comes home to itself, in and as absolute knowing. Here knowing no longer feels the need to go beyond itself; it is finally at home with itself, having absolved itself from all alienating otherness, for all otherness proves finally to be its own otherness. It even surpasses the *desire* for wisdom, as in previous philosophy, and becomes the possession of actual science, *Wissenschaft*. Previous philosophy was always *between* ignorance and wisdom; now there is no such between, because everything is between knowing and *itself*, in the circle of its own self-determination. In Hegel, after the old metaphysics, and the new critique, we are offered the new specu-lative philosophy which in post-transcendental form offers the totality of categories, each allegedly justified beyond critique, because having been radically critiqued by dialectic.

This dialectical way is carried on the labor of the negative to a media-tion of the equivocity, through the many determinate intelligibilities, all the way to fully self-determining knowing. While this triadic movement from indeterminate, through determination, to self-determination has a certain qualified truth, it is not fully true to the dimensions of the perplexity sug-gested above. Here too there is something that exceeds determination, some-thing also not to be described in the language of self-determining thought. If the latter takes itself to be the absolute measure of what is at issue, it suf-fers from the same bewitchment of the equivocity which it ostensibly claims

15. Hegel speaks of the journey of his *Phenomenology* as *der Weg des Zweifels*, the way of doubt, and we should note the reference to two, the double (*zwei*), something we also note when he describes the journey as a *Weg der Verzweiflung*, a pathway of despair. *Phenomenology of Spirit*, §78. There he also speaks of the *Phenomenology of Spirit* as a "self-accomplishing skepticism [*sich vollbringende Skeptizismus*]." See *Phän-omenologie des Geistes*, 67; *Phenomenology of Spirit*, §49.

to rationally mediate. It is within the Cave but has redefined its immanence as the whole, and hence is in an even worse position than those prisoners who know and grant with raw pain that they are still perplexed prisoners. The perplexity of the Cave has been dialectically domesticated: the Cave now is no Cave, because all that is there is (self-)determined as immanence at home with itself and beyond which there is nothing greater to be thought. Without perplexity we settle into a false home at whose hearth flickers the fire of self-determining immanence itself as its own counterfeit god.

Being Determinate: Wonder as Curiosity

I turn now to curiosity as a third modality of wondering, itself marked by its own doubleness, even despite its impatience with equivocity: the doubleness of being absolutely indispensable to the essential determination of being and cognition, and yet of always being tempted to run roughshod over the overdeterminacy of astonishment. The perplexity that can live on in think-ing as negation can be further dulled, even unto the death of wonder. If to be is to be determinate, here to be is nothing if it is not determinate. Being is nothing but determinacy and to be exhausted in the totality of all determi-nations. The "that it is at all" of given being is taken for granted rather than as granted. The danger here is that the necessity of determinacy becomes a turn away from the overdeterminacy, and hence into a kind of hostility to ontological astonishment. Curiosity about the determinacies of being risks becoming a configuration of wonder which leads to the annihilation of the wonder of being itself.

It is necessary to assert the *constitutive* role of curiosity in getting as univocal a grip as possible on the intelligibility of being, in addition to as precise an articulation as possible of that intelligibility. We cannot but be cu-rious, given that we are a desire to know the world around us and ourselves. When we are curious, our desire to know attends carefully to the details of things. Einstein admonishes us: "Never lose a holy curiosity." "Holy": it is as if the sacred companions the curiosity. In asking about curiosity we are also asking about the *worthiness* of knowledge, or at least certain kinds of claims to knowledge. In an adapted way, we are not far from Nietzsche's question about the value of truth. There may be claims to the true, claims bound to certain determinations of univocal truth, that carry us away from the spirit of truth, even as they are enabled at all by being carriers of the same spirit of truth.

We do think, of course, that sometimes curiosity can be excessive, that it can be addressed to unworthy things, that it can intrude too minutely into

things. We speak of a healthy curiosity but there also seems an inquisitiveness that is too intrusive and that is taken up with what does not properly
concern us. Curiosity, in a good sense, finds things interesting and surprising; its desire to know is open to the novel and strange.[16] It fastens on things
in their interesting determinacy. It is open to what is unfamiliar and odd.
The novel, the peculiar, the queer often draw the attention of the curious
mind. We sometimes judge an argument to be curious: departing from expected ways, it is marked by ingeniousness or too much subtlety. There are
those who are collectors of curiosities: in out of the way places, they are on
the lookout for things or people out of the ordinary. We are familiar also
with inquisitions, by contrast, where novelty is suspect. The inquisitor is
particular about details because there are details that show forth the unap-
proved. We might approve of the openness of the desire to know, but we are
more equivocal about something we think one has no right to know. There
is such a thing as *prying curiosity*. We deem it to intrude on what properly
does not concern it. Something is "off-limits" to the desire to know.

This double-edged character means that, *qua* wonder, curiosity is not
a pure porosity to what is true. What we are in the intimate recesses of our
being infiltrates our manners of being curious. There *can* be something
closer to the purer porosity, the reception of astonishment, the awakening of
perplexity. But it can also be the case that in the search for light something
darker surges forth. From secret intimate sources, the desire to know can be
marked by a *conatus essendi* that wills to overtake, subordinate, if not extir-
pate the porosity and patience that are intimate to our being. The doubleness
is not to be forgotten. There are formations of curiosity in which the will to
make being as univocally determinate as possible takes on an all-pervasive

16. I find it curious that the word "curiosity" became much more common in English usage, as well as in publications, and in a positive sense, sometime in the later
sixteenth century. See Ball, *Curiosity: How Science Became Interested in Everything*,
3–22. This was a time on the threshold of modernity, its turn to self, and the growing
objectification of being. That said, and balancing, I think of Charles Dickens's *The Old
Curiosity Shop*: the mystery of the human in the junk heap of flotsam and jetsam, bit
and baubles, trash and treasures, treasure hidden in the trash heap. Ball details some
premodern attitudes to curiosity that are more equivocal, if not deeply suspicious.
St. Augustine, *vitium curiositatis* [vice of curiosity]; St. Bernard, *primum vitium est
curiositas* [the first vice is curiosity]. By contrast with *vana curiositas* [vain curiosity]
only, Aquinas has an interesting discussion of *studiositas* [studiousness, the virtue of
attention] and the issue of directing curiosity and knowing (*Summa Theologiae* II-II,
qq. 166–67). See Pappin, "Directing Philosophy: Aquinas, Studiousness, and Modern
Curiosity." To the degree that my analysis of the modalities of wonder is ontological/
metaphysical, the equivocity of curiosity is not premodern, modern, or postmodern.
There is about it a saturated equivocity that is ontologically constitutive, and hence to
be found in, but not exhausted by, premodern, modern, and postmodern forms.

momentum that presents it as an irresistible power of its own. Its sources in the more original porosity are forgotten, as well as its salutary interplay with the other modalities of wonder in astonishment and perplexity.

If perplexity is a first-born child of primal astonishment, curiosity is a second-born. If astonishment is overdetermined, if perplexity mixes the overdeterminate and indeterminate, curiosity dominantly stresses the determinate. Often we think of wonder in this third modality as wrestling with "problems." This does make sense—the "It is!" of first astonishment turns into the "What is?" (indeed "What the hell is it?") of perplexity, turning now into the sober "What is it?" of curiosity.[17] When we pose the question in this third form, we are asking primarily about the *determinate* being there of beings, or the determinate forms or structures or processes. We have moved from ontological astonishment before being as given at all toward ontic regard concerning beings, their particular properties, their intricate patterns of developments, determinate formations, and so on. This movement into wonder as curiosity is essential to the genuine becoming of our mindfulness of what is. This follows from the fact that the overdeterminate is indeed saturated with determinations. It is not an indefiniteness empty of determinacy. The more determinate question "What is it?" turns us toward the rich, given intricacy of this, that, and the other. There can even be something of ontological and epistemic reverence in this turn to things. After all, it too participates derivatively in our original porosity to the astonishing givenness of being. Curiosity can release our sense of marvel at these given intricacies of things.

I am interested, however, in a certain understanding of curiosity which turns the teleology of wonder into a movement from the indeterminate to the determinate, and thence from determination to determination, all the way to the totality of determinations which are held to exhaust the whole. I connect Hegelian negativity with such a teleological movement from indeterminacy, through determination, to self-determining knowing. While this understanding is not quite to be identified with the view that being is simply determinate, nevertheless he reveals something essential, and I think questionable, about such a teleology: what seems mysterious in the initial indeterminacy is brought into the light of full intelligibility at the end of the unfolding, intelligibility determinable by knowing as self-determining. This is evident at the highest level of absolute spirit: art comes to an end when the enigma of the origin no longer retains anything secret; likewise, in the end religion safeguards no divine mystery that ultimately is too much for

---

17. If one tried to distill the three modalities and image them, it might look like this: Astonishment = !!; perplexity = !?; curiosity = ??.

the power of philosophical knowing.[18] Hegel's self-determination thus shares this crucial orientation with this understanding of the teleology of curiosity. This kind of curiosity negates the indeterminate, for this as such cannot be grasped, for only the determinate is thus graspable. Behind this grasping can operate a metaphysical *ressentiment* against anything in the ontological situation that exceeds its measure, a secret hatred of the overdeterminate. Equally all perplexity troubled by the "too-muchness" tends to be deemed an oppressive equivocity and as such no longer to be abided. There is no abiding with the mystery of given being. There is to be nothing abiding about the mystery of given being. If we look at the teleology of knowing in this way, we risk the eventual evacuation of spiritual seriousness not only in art and religion but also in philosophy. We come to suffer not simply from a dearth but from the death of ontological astonishment. For there is no room now for the *thaumazein* that recurs to the overdeterminacy in the never dispelled porosity of being. In claiming to fulfill its opening in fully self-determining knowing, it is no longer opened as a received porosity that sources the desire to know. Great art works, like religious reverence or awe, may offer us striking occasions of originating wonder that never dispel this ontological porosity but purge it of idolatrous lies. If original wonder is entirely impelled out of its initial hiddenness by determinative curiosity, the porosity is no longer kept open in philosophical mindfulness, and our ontological appreciation of the overdeterminacy of being withers. The wiser patience that waits on the renewal of first astonishment is betrayed.

18. On this in connection with art in relation to the teleological movement from symbolic, through classical, to romantic art, see my *Art, Origins, Otherness: Between Art and Philosophy*, ch. 3. In connection with religion, see *Hegel's God—A Counterfeit Double?*, and especially ch. 6 in relation to the idea of creation, which for Hegel is a "representation" that does not get to the true Hegelian concept, which is "creation" as God's own self-creation, God's self-determination. Creation is not the hyperbole of radical origination (see *God and the Between*, ch. 12), nor is the world as created the eventuation of finite being as given to be as other to the divine. The stress is not on such radical "coming to be" but first on becoming, then on self-becoming, indeed the self-becoming of God, and this following the teleological movement from indeterminacy, determination to self-determination. Just as there is no sense of hyperbolic giving to be, there is no sense of the baffling nothing out of which finite being is said to be given to be; there is determinate negation as the negativity immanent in the self-circling whole.

The Idolatry of Contracted Curiosity
and Its Limitless Self-Expansion

Is there a connection between self-determining thinking *qua* self-relating negativity and the dearth of ontological astonishment of scientism? I believe there is, and it has much to do with overlooking the otherness of given being in its overdeterminacy. I want to return to the equivocity of curiosity in its impulse to overcome equivocity. The contraction of wonder that curiosity can become, paradoxically, can also become infinitely self-expanding. Curiosity *qua* wonder is a form of the original porosity, but *qua* contraction to this, that, and the other determinacy, it is a limitation of the porosity. And yet this limitation, again due its being sourced in the original porosity, can become limitless.

Curiosity is a contraction of wonder, but *qua* wonder, it can be limitlessly interested in everything. This seems something to be scientifically lauded, but suppose we ask now not about the "what" concerning which we are curious but the "how" or the way of being curious about something, about everything. What looks like the same curiosity can be informed by a kind of secret love of the object or subject of curiosity, but it might also be informed by a kind of hatred of its resistant mystery or marvel. It might hate the latter because it does not yield univocally to determinate curiosity, and will never do so. "Questioning is the piety of thinking" (*Das Fragen ist die Frömmigkeit des Denkens*), as Heidegger deeply remarked. What he did not remark is that the same question can be posed in the modality of love of the true, or in the modality of aggression, even hatred. There are ways of questioning that lack reverence for the thing questioned. They are impious. They are an assault on the thing.

Perhaps we all have some experience of being so assaulted by a questioner like this. I can recall a person who would spit out questions at me at first meeting, and of course one retracted into oneself because of being so attacked by questioning. Look at the way lawyers ask questions: sometimes insinuatingly, sometimes seductively, and sometimes too with barely restrained violence. There are journalists like this too. It is as if what might be an act of wooing, perhaps in advance of love, has no finesse for the reserve of the intimate and instead strikes impatiently and directly, as if that were the way to truth. This approach does not leave the space of the between open. It does not allow anything to be itself and to reveal itself in its own more intimate ontological terms. There is no wooing, no seduction, but more the rape of an erotic assault; for, after all, wooing and foreplay are full of equivocity, while the direct movement to union may well be an impatient univocity without reverence for the reserve of the intimate.

I have suggested a wedding of determination and self-determination, insofar as the determination serves the self-determination as the secret end of the "project" of knowing as a whole. Whether this wedding is made in heaven is the question of scientism. Scientism, as it were, draws on the gift of heaven in the infinite openness and restlessness of the desire to know. The wedding of theory and practice: this wedding is the binding of the theoretical desire to know with the practical desire for know how. My question is: if this restlessness of curiosity is turned away from the wonder of being, is it not inclined to generate a kind of *counterfeit infinity*? In what form? Answer: in the form of an infinite restlessness that is not in search of the true object of its desire but of the infinitely repeated excitation of restlessness without end.

If this happens, we are in the business of producing the counterfeit double of our infinite desire. Call it: infinite desire to desire without the infinite. For without the infinite the desire mutates either into a venture into limitless exteriority or into self-circling self-excitation (perhaps these two are the same, at a deeper level). It is not infinite restlessness as a kind of intentional infinitude, which still is open to the answer(ing) of the truly and actually infinite. The latter would precipitate wonder in the modes of astonishment and perplexity, but wonder as univocally determining curiosity alone cannot stay for this revelation of the true infinite. Restlessness without end does reflect the intimate infinitude marking our desire to know but here its intention is not to be true to the real but (in the second option above) to realize itself as self-exciting without inherent limit. It is more like Hegel's self-sublating infinite rather than the human being's infinitude between its own self-transcendence and transcendence as other.

The wedding comes today perhaps dominantly in the marriage of science and technology. This is, in principle, a catholic marriage, catholic as having a bearing on the universal, as there is a globalized and institutionalized character to this now. But, in light of the self-expansion of contracted curiosity, one has to ask if this is a project of a universal that gives expression to our will to be the sovereign of the whole. Sovereign of the whole: proximately by drawing on the desire for the universal of the drive to scientific knowing; less proximately and indirectly, because energized by the drive for technological sovereignty in the form of the superiority of our project of unhindered self-determination (again both in a practical and theoretical respect). Is the universal of this globalized universal a counterfeit catholicism? One has to ask too, calling on Pascal: does not this universal reveal itself in the dominance of the *esprit de géométrie*, not the *esprit de finesse*? It is not the intimate universal where ontological intimacy does not eschew geometry but ties it back to the gift of originating astonishment.

This wedding draws from the original porosity and wonder as astonishment but it determines the porosity in terms of human power, which
means the mutation of the porosity, perhaps even denial of it. This leads
to the contraction of the opening of curiosity in the very claim to fulfill it
further and more fully. Hence the counterfeit infinity.

Is something of this reflected in economics as under the sign of will to
power, not under the *oikos*[19] of the home on earth we are to build for ourselves? Will the counterfeit curiosity destroy all domiciles? For after all, in
the *domus*, the domicile of the *oikos*, must not intimate life be allowed to be,
allowed to be in receptivity to its deepest originating ontological sources?

The matter touches on death and life, especially in their ultimate extremities: war and health, the one dealing with our killing machines that
emerge as the expression of our endeavor to be, the other dealing with the
suffering that is unavoidable at the extremity of that endeavor to be. One
hates our porosity and *passio essendi*; the other tries to keep at bay the *passio essendi* that comes back again and again, in suffering, and in sickness,
and in death, as the ultimate return of the patience of being, now in the
shape of not-being, and at the threshold of the mystery that is our entry
into posthumous porosity. Look at how for a century at least so much of
the scientific enterprise has been conducted under the patronage of either
the military-industrial complex or the pharmaceutical industry. These are
mass formations of the marriage of theoretical and practical curiosity. Astonishment and perplexity get in the way of scientistic determinability and
superimposing self-determination.

I wonder also if this contraction of curiosity in its limitless self-expansion is connected with our determination of the whole of being in the light
of serviceable disposability. A totalizing of the useful produces the reign of
serviceable disposability: things are made to serve our instrumental desire,
and hence be serviceable, but when used, they are used up, and are hence
disposable. This point has some implications for wonder in the modalities of
astonishment and perplexity. There is something about them that is deeply
and intimately beyond serviceable disposability. One aspect of their indispensable service is precisely the way they open up for us the dimension of
the given *qua* given, prior to use, and beyond use. There is a transcendence
to use of what strikes us into astonishment and perplexity. What is there is
not there to be used up. Art can sometimes refresh original astonishment
and perplexity.

---

19. Meaning: house. The eco- in economics (lit. "household management") is derived from *oikos*.

What of our being religious? Our being religious is pluriform, but if we consider the idea of the superior other, granted in light of divine transcendence, the space between divinity and us will not be best served by limitless curiosity, oriented to determination or self-determination. If wonder, especially in the modalities of astonishment and perplexity prepare and keep purged our intimate porosity to the divine as the superior other, then all our efforts at completing the project of seamless self-determination, via our determination of being as other, are not only prone to failure, but are already themselves expressions of a hubris on our part. The project to succeed thus is already a failure. It is true that if one were familiar with the longer religious tradition, this project itself would have difficulty avoiding the name of pride, the primal sin. Our turn to ourselves as agents of complete self-determination—and surely this is part of the agenda of modernity—cannot but look with suspicion on any claims of superior otherness. As a consequence, our relation to God as such a transcendent other must be itself relativized, called into question, perhaps even liquidated, if its resistance to our autonomy is ineradicable. In so many words, this is curiously reminiscent of the project of the serpent in Eden and its seductive promise: you too will become as gods. There is a certain modernity that is in the business of trying to prove the serpent right.

If there is a secret form of tyrannical will to power behind the scientistic desire to reconfigure all of being according to its dictates of a certain globalizing univocity, the serpent of counterfeit wisdom is its secret seducer to the false whole of reconstructed creation. Such a reconstruction has the tendency to fall into de-creation rather than creation. For it is always parasitical on the gift of creation which it can never acknowledge as such, since its project of the perfect future is just fueled by the energy of this refusal of the given as such. Its "creation" is de-creation: another voiding of being. To convert from refusal to acknowledgment would be to allow the offer of the more original porosity to give itself again, with all the painful surprise of marvel, long since eschewed as the brand of one's humiliation by the superior other. Humility before the gift of being would have to come again in place of the guarded protection against the gift of goodness that is perversely experienced as one's own humiliation. The original wonder gifts us with the gift of being able to receive gifts. If it is contracted or mutilated in scientistic curiosity, then we cannot accept any gift, we cannot accept anything as given, for it is not given on our terms, and hence it is a burden to be dismissed, incompatible with our status as self-sublating gods.

Is it surprising that instead of the perfected earth we have the despoliation of nature and the oncoming night of ecological catastrophe? We are making gardens of the earth into fetid dumps. And it is our violent intrusion into everything, with no reverence that lets things be as they are, that loves things even as they are, that blights the goodness of creation. The blight is like the bleaching of the coral reefs. Almost suddenly, the blank counterfeit is revealed, and the habitat of life is made a dead zone for creatures who would otherwise enjoy the bounty of the maternal sea.

Both the will to reconfigure the whole, and the ontological loss that threatens more and more to come as the destiny we impose on immanence, are long-term consequences of the infinite self-expansion of the contraction of curiosity: its mutation not only of the original wonder but in that mutation loss of original love of the given mystery of being. In scientistic curiosity we see then an exploitation of the porosity but in a contraction of its purer openness. In the exploitation of the purer porosity there is the loss of porosity, and its being overtaken by an irreverent will to power, especially in the useful form of serviceable disposability. A counterfeit infinity of the desire to know: does it really desire to know, or does it want to be number one, be the alpha being, in order to be the omega being? We can never be alpha beings because we are always seconds: we come *after*, because the alpha gives us to be, gives us to be as porosity to being other than ourselves. The true alpha being gives being to be, and not in the domineering form of imposing sovereignty—rather, in the gift of agapeic letting be: from surplus, not from a lack that has to counterfeit infinity to feel immanently in itself its own absoluteness. The true alpha is already fully itself: superplus, overdeterminate, the generous infinity. Fully itself: even when it agapeically makes way by giving a way, and lets be kenotically, as if it were nothing in the way. All counterfeits come down to wanting to be it, but not in the reverent modality of wanting to be like God but in the irreverent usurpation of wanting to be god. The essence of scientism is thus at core religious. Its "project" is the construction as the graven image of our time: self-determined, self-enclosed immanence as the changeling of true infinitude. As secreting a counterfeit infinity its essence is idolatry.

Bibliography

Ball, Philip. *Curiosity: How Science Became Interested in Everything*. Chicago: University of Chicago Press, 2013.

Burke, Edmund. *A Philosophical Enquiry into the Origin of our Ideas of the Sublime and the Beautiful*. Edited by Adam Phillips. Oxford: Oxford University Press, 1990.

Coleridge, Samuel Taylor. *Samuel Taylor Coleridge, The Major Works*. Edited by H. J. Jackson. Oxford: Oxford University Press, 2000.

Crews, Frederick. *Freud: The Making of an Illusion*. London: Profile, 2017.

Desmond, William. *Art, Origins, Otherness: Between Art and Philosophy*. Albany, NY: SUNY Press, 2003.

———. *God and the Between*. Oxford: Blackwell, 2008.

———. *Hegel's God—A Counterfeit Double?* Aldershot, UK: Ashgate, 2003.

———. *The Intimate Strangeness of Being: Metaphysics after Dialectic*. Washington, DC: Catholic University of America Press, 2012.

Harari, Yuval Noah. *Homo Deus: A Brief History of Tomorrow*. London: Harvill Secker, 2015.

Hegel, G. W. F. *Phänomenologie des Geistes*. Hamburg: Meiner, 1952.

———. *Phenomenology of Spirit*. Translated by A. V. Miller. Oxford: Clarendon, 1977.

Heidegger, Martin. *What Is Called Thinking?* Translated by J. Glenn Gray. New York: Harper and Row, 1968.

Pappin, Gladden J. "Directing Philosophy: Aquinas, Studiousness, and Modern Curiosity." *Review of Metaphysics* 68.2 (2014) 313–46.

Wordsworth, William. *William Wordsworth*. Edited by Stephen Gill. Oxford: Oxford University Press, 1984.

PART TWO

2

Preparing to Paint the Virgin's Robe

Spike Bucklow

As a practicing scientist, I would like to provide an empirical, bottom-up, response to William Desmond's philosophical, top-down, analysis of curiosity, perplexity, and astonishment in science. Curiosity may indeed be a scientist's private motivation but it is not usually acknowledged. The rhetoric of scientific funding applications usually stresses conformity to, and the potential to extend, existing programs of research, often additionally requiring a direct or indirect impact on saving lives, taking lives, and otherwise making money or, alternately, contributing to nation-building identity and status. The encultured scientist repackages their curiosity accordingly and enculturation is such that expression of personal motives—especially those involving the ineffable, such as beauty—is to be expected, if at all, only after career progression has peaked.[1]

I have a background in chemistry and experience in the material aspects of cultural artefacts. My activities fall under the umbrella of heritage science and some can be justified as informing the conservation profession and ensuring the continued physical integrity of cultural objects. However, the research that gives me most pleasure—rediscovering how historic painters worked—has no such obvious utility (as the historian Barthold Niebuhr said, "He who calls what has vanished back into being enjoys a bliss like that of creating").[2] Such research forms part of technical art history, a relatively new academic domain characterized by many case studies and few

1. Hardy, *Mathematician's Apology*, passim.
2. Secord, *Knowledge in Transit*, 672.

overarching theories. The following draws upon case studies that will together, I hope, illuminate aspects of the relationship between modern science and astonishment.

Art history occasionally contains appreciative statements about its
objects of study. For example, the face of Christ in Raphael's final painting
has been compared to a "theophanic vision."[3] However, most art history
approaches paintings analytically. For example, Raphael's works have been
deconstructed into those consciously sought influences that can be univocally determined, such as Perugino's rendering of space, Andrea del Sarto's
brushwork, Sebastiano del Piombo's color, etc.[4] An artist like Raphael would
have seen no great disjunction between art and science but, even after science parted company with art, individual scientists have still been attracted
to the study of art. Michael Faraday, for example, analyzed artworks' material compositions.[5] Artworks might appeal as objects of study because they
are particularly accessible products of Desmond's "original porosity."[6] This
may also account for their cultural status as objects of power.[7] (For example,
Faraday's analysis was of the paint on looted sculptures in the British Museum and a UK Act of Parliament currently prevents their return to Greece.)

Science's first major contribution to art history was in helping to
determine the univocal identity of instances of one class of power-laden
object, the so-called Old Master painting. In the nineteenth century, Italian
paintings were finding new homes in northern Europe and the New World,
contributing to nation-building identity and status. Controlling the accel-
erating export of cultural heritage was complicated by the fact that most
attributions were wildly optimistic and practically every painting was allegedly by a big name like Raphael. To address this problem, an Italian politician developed a system of identifying artists by focusing on those parts
of paintings that viewers generally overlooked and artists repeated across
many paintings—ears, hands, etc.—using them like signatures. Giovanni
Morelli thus deconstructed the image and sought out the painter's passive, unconsciously produced, effects. This forensic approach proved very
successful, but its apparent demystification of paintings was contentious.

3. Britton, "Raphael and the Bad Humours," 192.

4. Williamson, "The Concept of Grace," 318–19.

5. Hamilton, "Report to the Committee," 105–7.

6. Desmond, *The Voiding of Being*, 99–100, 109, 118, 120, 122, 124 [18, 19, 27, 35,
37, 39, 40].

7. DeMarrais and Robb, "Art Makes Society," 3–22.

Morelli himself stressed the role of an ineffable "quality" as the final arbiter in attribution.[8]

The method's success spread beyond paintings and found expression in the work of Sir Arthur Conan Doyle who, like Morelli, had once studied comparative anatomy. Conan Doyle's hero, the fictional detective Sherlock Holmes, deconstructed crime scenes and sought out the perpetrator's passive, unconsciously produced, effects, like footprints. Conan Doyle even acknowledged his debt to Morelli in stories that featured clues like severed ears.[9] Like Morelli's method, Conan Doyle's hero was very successful, to the extent that his creator felt trapped by, and compelled to kill off, his creation. The public, however, demanded the detective's return and Conan Doyle relented with *The Hound of the Baskervilles*, now widely accepted as the best Holmes novel. However, close reading of that novel shows how Conan Doyle makes Holmes' repeated blunders responsible for a largely unrecognized miscarriage of justice.[10] Conan Doyle's subtle literary joke might suggest that, like Morelli, he was aware of the limits of univocal interpretations of supposedly objective signs.

If science's first contribution to art history involved interpretation of visual data, its current contribution mainly involves material data, following in Faraday's footsteps. It also follows the Morellian focus on passive, unconsciously produced aspects of paintings, this time premised on habitual ingredient selection, rather than habitual gestures recorded by loaded paintbrushes. The objective signs used by heritage science and technical art history range from lead isotope ratios or the presence of trace elements to the irregular spacing of growth rings in wooden panels or warp threads in bolts of hand-woven canvas. Through changing patterns of global trade, such signs can, to greater or lesser extents, act as markers for the date and location of a painting's origins. Assuming the painting's creator is unaware of the scientist's chosen focus—or, if aware, is unable to control their ingredients' origins—then such methods can identify anachronistic objects. For example, a painting that is claimed to be by Raphael would probably not contain a lead pigment with an isotope ratio corresponding to that found in a New World lead-containing mineral because early-sixteenth-century trans-Atlantic trade in lead was negligible.

The findings of such scientific analysis are conjectural and depend on many cultural variables, including historic artists' studio practices. Raphael, for example, employed studio assistants who routinely completed sections

8. Morelli, *Italian Painters*, passim.

9. Ginzburg and Davin, "Morelli, Freud and Sherlock Holmes," 5–36.

10. Bayard, *Sherlock Holmes was Wrong*, passim.

of his paintings—like Perino del Vaga for architecture or Giovanni de Udine for animals—and Giulio Romano even finished whole "Raphael" paintings.[11] Science is therefore only capable of a univocal identification of a painting's status under extremely rare circumstances, such as the Vermeer purchased by Hermann Goering that turned out to have been painted with oil incorporating the ingredients of the twentieth-century resin, Bakelite.[12]

Any ambition to identify a Raphael on the basis of scientific analysis is arguably an example of scientism. Acknowledging, as Desmond says, that "art comes to an end when the enigma of the origin no longer retains anything secret," my own scientific analysis does not pursue paintings' identities.[13] Rather, I try to use scientific analysis to explore how "[a]rt can sometimes refresh original astonishment and perplexity."[14] Such an approach, however, marginalizes my work since the more highly cited heritage science journals tend to privilege research with univocal outcomes.

Up to the seventeenth century, most painters learned how to make paint when young and as their careers progressed their paint was made—to their specification—by studio assistants. Part of the painter's skill therefore lay in preparing their materials, and their widely recognized artistic skills rested upon less-recognized craft skills. A surviving painting is a physical record of craft skills, which are practical expressions of knowledge that we would now call physics and chemistry.

My research started by reading artists' recipes in historic treatises, through the lens of the four Aristotelian elements, and then practically re-enacting those recipes. Where procedures were well-specified, they usually proved efficacious. Of course, that does not prove the underlying Aristotelian theories were correct—any more than modern science's theories are—since, as St. Thomas Aquinas said, perhaps "appearances can be saved" in other ways.[15] Well-specified historic recipes usually involved univocally identifiable ingredients like oil and water, used to separate the different colored minerals in lapis lazuli, or sulphur and mercury, joined together to make vermillion.[16] My attention has since shifted towards ill-specified recipes, as accessible examples of processes that involve embodied or tacit knowledge.[17]

11. Vasari, Lives', vi, 145.

12. Lopez, Man Who Made Vermeer, passim.

13. Desmond, The Voiding of Being, 119 [36].

14. Desmond, The Voiding of Being, 123 [39].

15. Blanchette, Perfection of the Universe, 11.

16. Bucklow, Alchemy of Paint, 43–108.

17. Polanyi, Tacit Dimension, passim.

In his c. 1400 treatise, the Italian painter Cennini wrote about the preparation of the blue pigment, azurite, saying only that it was "very scornful of the stone" and should be "worked up" with water. He was slightly more forthcoming about malachite, a related green pigment (both are mineral copper carbonates). There, he recommended a "light touch" since too much grinding gives it a "dingy and ashy color." He then said you should stir the powder in water, let it stand and pour off the water, saying "wash it this way two or three times and it will be still more beautiful."[18] Azurite recipes in other treatises provide little or no more additional detail.

I can discern no contemporary theoretical underpinning for this procedure and it was probably inspired by the observation of color changes upon grinding pigment or washing brushes. (How many of us can claim to be "porous" and "open" to inspiration while routinely preparing for, or tidying-up after, what we consider to be our work?) I think the recipe was simply a response to perplexity. After all, why should azurite scorn the stone and lose its color when most other pigments retain their color upon grinding? Cennini even said of vermilion that "if you were to grind it every day for twenty years it would still be better and more perfect."[19]

Modern science provides partial explanations for the behaviors of materials that painters observed and harnessed over seven hundred years ago. Grinding solid minerals produces particles of various sizes, generally coarser with a "light touch" and finer upon more protracted grinding. For reasons that are still not clear to modern science, in azurite (and malachite) the very fine dust-like particles scatter white light whereas the larger particles reflect a deep blue (or green). A "dingy and ashy color" results from the presence of very small particles and the "still more beautiful" color results from removing those dust-like particles. The recipes therefore allude to the separation of different sizes of particles of the same material.

Re-enactment of the recipe is complicated by lack of detail. Is there a cup of water or a bucket full, is there a lot of powder in the water or not much, how long does each stage take? Other recipes specify timing by chanting a number of *Pater Nosters* or *Ave Marias*, but Cennini only says let the water stand for an "hour." (Even a short mid-winter hour—one twelfth of the period between dawn and dusk—felt excessive to me, perhaps betraying my modern focus on productivity.) However, just trying the recipe, even in the absence of someone who could demonstrate, quickly indicated its approximate parameters. It was as if the world—or, at least, azurite—is not "indifferent" but is instead "an intelligent accomplice, a sympathetic

18. Cennini, *Craftsman's Handbook*, 35–36, 31–32.
19. Cennini, *Craftsman's Handbook*, 24.

partner, aiding and abetting the mind's efforts to understand its structure and to define its laws."[20] The following interpretation of the recipe has been validated by reference to microscopic samples of azurite taken from medieval paintings. In this, I have followed Morelli and Holmes by treating historic paints' physical compositions as footprints left by artists'—or their assistants'—habitual gestures.

First, what is a "light touch"? Grinding too little results in a difficult-to-use gritty paint, so significant grinding is required, despite the alarmingly rapid descent into a "dingy and ashy" color. The "light touch" is relative to other pigments that, like vermilion, get "better and more perfect" with extended grinding, and so assumes a knowledge of the processes' original context. Yet the process itself also provides clues. When you start grinding the brilliant blue rock fragments on a granite slab they make a hideous screeching sound, like chalk on a blackboard. But soon the shrieking subsides into the sound of flowing gravel, then into a hiss and finally it starts singing to you. The sound's shifting quality is an expression of the range of sizes of particles between the muller and slab. With an attuned ear, the azurite will tell you when it has achieved the appropriate mixture of fine and course particles.

Second, what about the scale? Modern material-use is typically profligate, so it is a shock to find that efficiently washing azurite involves around a thimble-full or shot-glass of water. (Pictorial and archaeological evidence suggests that discarded mussel and oyster shells were popular vessels for processing pigments.) Acquiring enough azurite to paint the Virgin's robe for an altarpiece takes hundreds or even thousands of separate batches, each washed "two or three times." In each batch, only a tiny pinch of powder is processed, similar to the amount of salt you might use to season your food.

Next, how to "stir"? This seems best done with a finger, rubbing the powder against the smooth mother-of-pearl—or equivalent—of your chosen vessel. This serves two purposes. First, clumps of particles are broken up, as the fine dust-like colorless particles are initially stuck to the bigger colored ones by the static generated in grinding. Then, the free-floating dust and bigger particles are both lifted up into the water and partially kept in suspension by gently swirling the vessel.

Letting the water "stand" takes advantage of the fact that suspended particles fall through water at different rates and—in this case, where they are all the same material—the speed depends on their size. Bigger ones fall fast whilst smaller ones remain suspended (not unlike the biblical wheat and chaff). Rapidly, a deep blue layer is deposited but, over time, the pale

20. Jacks, *Alchemy of Thought*, 28.

matter follows. Of course, there is not a binary distinction but a continuum of particle sizes and shades, so exactly how long you leave it is a judgement call. Letting it stand longer gives you more blue, but the mixture is paler. A briefer respite yields a smaller quantity of a deeper blue.

Pouring decants the water that disproportionately carries the smaller, paler particles and it involves another judgement call. Decanting less will contaminate the deposited beep blue with pale blue whilst decanting more risks reducing the yield of deep blue. Consequently, both the material that remains in the mussel shell and the material that was decanted from it are both re-washed, two, three (or more) times. Eventually, all water is decanted from all containers to give many separate shades of dry blue powder. All are kept and used, the paler grades being used for painting skies and deeper grades in the Virgin's robe, for example.

In practice, grinding and cleaning azurite can be therapeutic, even without chanting *Pater Nosters* or *Ave Marias*. It can even prove addictive; in one workshop I ran, postgraduate students skipped their coffee breaks to continue washing. The easy, gentle, repetitive, and almost hypnotic hand movements are rewarded with increasingly beautiful, velvety-rich blues that glitter like the starry midnight sky. The sense of achievement can be profoundly satisfying and after another workshop, without seeking permission, an internationally respected history of science professor went round the studio pocketing everybody's richest blues, her inner magpie awoken by the fresh pigment's beauty.

To anyone who has washed azurite, it is obvious that the astonishingly rich blues in medieval illuminated manuscripts are made with paint containing no light-scattering pigment dust. For those who have not experienced the process, the lack of small particles in deep blue paint can be confirmed by examination with a scanning electron microscope. Upon realization, I was astonished by artists' ability to routinely create powders with particles of between 1 and 10 micrometers and no particles of less than 1 micrometer. With tolerances of around one-thousandth of a millimeter—close to the wavelength of light—this was highly sophisticated material processing achieved with a big granite slab, a dome-topped, flat-bottomed hand-sized piece of granite, some water, and a bag of mussel shells.

My rationalization of the process—particle disaggregation, differential sedimentation, etc.—is indebted to modern science, and the process is amenable to scientific study—correlation of audible frequencies with particle size distributions, etc.—but the poetics I have alluded to are beyond science. In fact, they were purposefully written-out of modern science at its very inception. Early fellows of the Royal Society aimed to appropriate the knowledge accumulated within medieval craft traditions—many papers in the early *Philosophical Transactions* feature pigment preparation—and they were guided by, amongst others, Francis Bacon. In his *Advancement of Learning* he recommended studying "arts" defined by (univocally identifiable) materials, saying "[t]hose which consist principally in the subtle motion of the hands . . . are less use."[21] Thus, azurite processing's slow, steady epicycles with a heavy granite muller, its gentle rubbing against mother-of-pearl, its repeated swirling and delicate pouring were banished from science. With them went the lived experience that there is no univocally correct way of doing—or fully understanding—anything that possesses significant complexity. Even an apparently simple, menial activity like grinding and washing rock can offer a window onto the ineffable. (Of course, in practice, experiments in the modern scientific laboratory can also involve craft skills, but—like personal motivations—these are not generally acknowledged.)

The limitations of modern—quantitative—scientific knowledge were built-in centuries ago and current scientistic tendencies hinder modern science's reintegration into the larger—qualitative—truth categories and belief systems that we have always used and enjoyed. Such belief systems include those of cultures and sub-cultures (including craft traditions) that do not dismiss the sensations of bodies in motion or colors and sounds, etc. Those cultures and belief systems are enriched by their respect for the curiosity, perplexity, and astonishment that such sensations can evoke.

Bibliography

Bayard, Pierre. *Sherlock Holmes Was Wrong*. Translated by Charlotte Mandell. London: Bloomsbury, 2008.

Blanchette, Olivia. *The Perfection of the Universe*. University Park, PA: Penn State University Press, 1992.

Britton, Piers D. G. "Raphael and the Bad Humours of Painters." *Renaissance Studies* 22.2 (2008) 174–96.

Bucklow, Spike. *The Alchemy of Paint*. London: Marion Boyars, 2009.

Cennini, Cennino. *The Craftsman's Handbook*. Translated by David V. Thompson. New York: Dover, 1960.

DeMarrais, Elizabeth, and John Robb. "Art Makes Society." *World Art* 3.1 (2013) 3–22.

Desmond, William. *The Voiding of Being*. Washington, DC: Catholic University of America Press, 2020.

Ginzburg, Carlo, and Anna Davin. "Morelli, Freud and Sherlock Holmes: Clues and the Scientific Method." *History Workshop* 9 (1980) 5–36.

Hamilton, W. R. "Report to the Committee Appointed to Examine the Elgin Marbles." *Transactions of the Royal Institute of British Architects of London* 1.2 (1842) 102–8.

Hardy, Godfrey H. *A Mathematician's Apology*. Cambridge: Cambridge University Press, 2013.

21. Houghton, "History of Trades," 38.

Houghton, Walter E. "The History of Trades." *Journal of the History of Ideas* 2.1 (1941) 33–60.

Jacks, Lawrence P. *The Alchemy of Thought*. London: Williams and Norgate, 1910.

Lopez, Jonathon. *The Man Who Made Vermeer*. Orlando: Harcourt, 2008.

Morelli, Giovanni. *Italian Painters: Critical Studies of Their Works*. Translated by Constance J. Ffoulkes. 2 vols., London: Murray, 1892–93.

Polanyi, Michael. *The Tacit Dimension*. Chicago: University of Chicago Press, 2009.

Secord, James A. "Knowledge in Transit." *Isis* 95 (2004) 654–72.

Vasari, Giorgio. *Lives of the Most Eminent Painters, Sculptors and Architects*. Translated by Gaston du C. De Vere. London: Macmillan, 1912–15.

Williamson, Edward. "The Concept of Grace in Raphael and Castiglione." *Italica* 24.4 (1947) 318–19.

3

Cultivating Wonder

Steven Knepper[1]

MY YOUNGEST DAUGHTER STOPPED, pointed, and whispered a poem: "Watch out—a tornado storm of bugs." I had not noticed the gnats funneling in our yard before her warning, but for a long moment we stood in the muggy evening haze and marveled at them. My four-year-old's epiphany revealed the swarm's chaotic order.

William Desmond is right that children tend to be more open than adults to astonished wonder at the richness of being, a wonder saturated in the strangeness that anything exists at all. If we let them, they can teach us to be more "childlike" in this sense. They can awaken us once more to neglected wonders.

Unfortunately for both children and adults, however, this is only part of the story. As Desmond shows, children are more open to wonder because they are (usually) more *porous* than adults. They lack tunnel-vision focus on long-term projects. They are not yet inured to the ways of the world. As a result, children are often more open to astonished wonder. But they are also more likely to be perplexed and even horrified at the everyday.[2] The same daughter who wondered at the gnats is terrified of bees—to the point that she sometimes has nightmares about them. Children are more open to the influence of others, as well, especially the influence of peers and parents.

1. A few passages and ideas in this piece first appeared in Steven Knepper, "Rediscovering Wonder" in *Plough Quarterly* 7 (Winter 2016): 62–64. I am grateful for the editors' permission to revisit them here.

2. Alexandra Romanyshyn illuminates the relationship between childhood, astonishment, and perplexity in her recent essay "Metaxology and Environmental Ethics."

Children are, as the cliché goes, impressionable.[3] In any good-sized group of young children, for instance, you will likely find at least one thoroughgoing cynic or skeptic or know-it-all who isn't impressed by much of anything. Such children often model their stance on an older sibling or parent. They may be children in age, but they no longer have a childlike openness to wonder. In mimicking the older and the jaded, they become so themselves.

If children are to maintain their "inborn sense of wonder," then, they will need "the companionship of at least one adult who can share it." They will need at least one adult open to "rediscovering . . . the joy, excitement and mystery of the world we live in."[4] These words are from the pioneering nature writer Rachel Carson's *The Sense of Wonder*, which describes how she introduced her young nephew to the Maine woods and seashores. Her point about the need for sympathetic adults may be even more important today, over half a century after Carson's book was published. A child's sense of wonder has probably always been fragile, easily fading with the transition into adulthood. But contemporary Western culture, with its pervasive achievement-seeking, consumerism, irony, and digital mediation—with what Desmond calls its "ethos of serviceable disposability"—can be particularly inhospitable.[5] In such a culture, as Desmond suggests, the richer wonder of the child is often directed into an instrumentalizing curiosity.

Looking back on my past, I realize the truth in Carson's exhortation. I was fortunate to spend my childhood in the care of adults who cultivated my openness to wonder. I grew up on a small dairy farm in Pennsylvania. There were pastures and woods to explore; I had meaningful work; and I was part of a warm community that contemplated the landscape, the weather, the flora, and the fauna. Every time a deer stepped out of a cornfield, every time a woodpecker landed on a tree, the adult with me would point it out, and we would pause together to watch it. Stories of animal sightings were often traded during visits with neighbors or extended family.

Part of this grew out of the necessities of rural culture. My family, like many in our area, raised or hunted much of its food. We scorned poachers who cut the horns off a deer and left the meat to rot. I spent many long, cold November days with my father in a deer stand. Necessity does not account

---

3. We could also turn to Dostoevsky's *The Idiot* for this lesson. Marie becomes a social pariah after her lover abandons her. The children, porous to the town's self-righteous indignation and scapegoating, torment her mercilessly until Prince Myshkin stirs their kindness and compassion. He guides them to an agapeic receptivity and service.

4. Carson, *The Sense of Wonder*, 45.

5. Desmond offers a developed account of the ethos of serviceable disposability in *Ethics and the Between*, 415–41.

for all, however. We would watch the deer come out of the woods on a summer evening not so much to scout for that fall's hunt as for the sheer enjoyment of it. The long stretches of patient waiting in nature encouraged what
Desmond calls "agapeic mind," a kind of "noninstrumental vigilance" receptive to the world.[6] An older cousin took me on many adventures. In the
spring we would wade through a sea of mayapples, hunting for mushrooms.
He taught me to identify trees, since morels tend to grow around elm and
ash trees. As evening descended, we would listen for wild turkeys flying up
to roost, and he would hoot like an owl to see if he could get them to gobble.
Once, when I was very young, he took me up a steep hollow at twilight to
watch a huge colony of bats fly out into a meadow to feed on insects.

I had a remarkable teacher for two years in grade school who built
upon the ecological affordances in our rural culture and challenged some
of its complacencies. She had us plant trees on the edge of the playground
and vegetable seeds on the windowsill. We had a succession of class "pets"
in terrariums and aquariums that we would observe and then release back
into their natural habitat. She encouraged us to turn her classroom into a
sort of makeshift nature museum, bringing in snakeskins, turtle shells, old
beehives, antler sheds, fossils.

My family life was also saturated in a religious ethos, one that mingled
awe for the natural and the supernatural. We walked to the Methodist church
on many Sundays, often stopping to peer into the creek on the way. The
words of blessing and burden, of gratitude and thanksgiving, came easily.
The "it is good" of creation and Adam's curse of toil and struggle illumined
the farming life, uniting the cosmic and the particular.

To be sure, this was no golden world where humans lived in harmony
with each other and nature. A child growing up on a working farm learns
early about death and suffering: the stillborn calf, the crops withered by
drought, the fawn caught in a haybine, the periodic collapse of milk prices,
the frustrations and anger that attend these. (Again, the child is more open
to astonishment, perplexity, *and* horror.) Nor did I grow up among impec-
cable stewards. There were hardly any birds of prey in my area during my
childhood, for instance, because generations of farmers shot them on sight.

While I did not grow up in a golden age, though, I cannot make sense
of my childhood without a sort of fall narrative. The Internet arrived in
my area when I was in the middle grades. My parents were excited about
the educational opportunities it would afford. (A few years earlier they had
stretched their budget to buy my sister and me a set of encyclopedias that
the Internet relegated to a dusty shelf.) They were not necessarily wrong

6. Desmond, *Perplexity and Ultimacy*, 123.

about this, but I quickly began to spend more time absorbed in the screen. I lost a good deal of my openness to the place and people where I lived, a good deal of my contemplative patience. I was well primed to abandon the ethos of my upbringing in my college years. My curiosity was increasingly shorn from its deeper roots in astonishment, and my life became shallower and more selfish and ultimately a good deal less happy as a result.

I didn't fully appreciate my childhood guides until I had a daughter of my own and began trying, often less skillfully, to recreate some of their lessons for her. Thinkers like Desmond and Carson helped me to get a better sense of why it was worth the effort, even if I wasn't as knowledgeable as my farming parents or naturalist teacher or woodsman cousin. The guidance is less about facts, about determinate knowledge, than it is about holistic experience. Carson stresses that she did not conduct detailed nature lessons on her walks with her nephew Roger. The emphasis was on observation and experience. They were "just going through the woods in the spirit of two friends on an expedition of exciting discovery."[7] Nonetheless, Carson notes that Roger learned much about animals and plants along the way, more than he would have from explicit educational "drilling." To parents worried that they do not know enough to take their children on such walks, Carson says "it is not half so important to *know* as to *feel*. If facts are the seeds that later produce knowledge and wisdom, then the emotions and the impressions of the senses are the fertile soil in which the seeds must grow."[8] Carson the scientist values scientific knowledge, but as a conservationist, she knows that experiences of awe, gratitude, and beauty help us recognize the worth of things.[9] And as an aunt, she knows that such experiences bring meaning into a life. They are not strictly subjective experiences. They are enriching encounters with a rich world. Carson's non-reductive science is thus far removed from scientism.

Desmond might say here that Carson cultivates the roots of curiosity in astonishment. Recall that Desmond's three modalities of wonder (astonishment, perplexity, and curiosity) are not resolutely separate from one another. Nor must they follow the familiar progression in which initial astonishment gives way to stark curiosity. Desmond notes that "there is a fluidity of passage of one into the other, indeed of permeability and mingling of one and the other."[10] This is one of the great strengths of Desmond's

7. Carson, *The Sense of Wonder*, 18.

8. Carson, *The Sense of Wonder*, 45.

9. On the relationship between ethics and aesthetics, see *Ethics and the Between*, especially 177–91.

10. Desmond, "The Dearth of Astonishment," 106–7.

account of wonder. It allows for descriptive nuance, for supple attentiveness to the differing emphases in various experiences. It also makes his account of wonder hopeful. Curiosity *can* remain rooted in astonishment. It does in Carson's writings, and it does in the work of many great scientists.

It is a challenge today, given the wider cultural ethos, to keep our children rooted in such astonishment. My own youthful experience of the deracinating effect of screens has been compounded and expanded through smartphones and tablets, through the omnimediation of the digital. Now more than ever, we need the kind of guides that Carson calls for. We need practices and pedagogies that cultivate an openness to astonished wonder. This is especially true in science, which has such authority in our time. We need a non-reductive science that recognizes the richness of our embodied experience of the world, that remains rooted in ontological astonishment. Wherever possible, children need to spend time exploring, playing in, and patiently observing nature. They need to tramp through the woods and get soaked in streams. They need to experience the *otherness* of nature, and they need adults who affirm and share their wonder at the aesthetic richness and singular intricacies and astonishing strangeness of it all.

The pervading cultural common sense is a reductionism in which the richness and singularity of things might be fun to look at it, but their reality is found at deeper levels—in cells and systems and the quantifiable aspects of being. These deeper levels are indeed important, but we only reach them by peeling away those aspects of experience that are charged with qualitative worth.[11] We risk losing the worth in the dissection. Children need adults— and curricula—that move beyond such reductionism. They need adults that affirm that the "too-muchness" of the world is not an epiphenomenon or a subjective fantasy but a reality. As Desmond has been pointing out for decades, we need a new metaphysics of "overdetermined" being. Metaphysics, in the broad sense of a set of assumptions about being, is unavoidable. We often unreflectively adopt metaphysical assumptions from our broader culture. And when that culture is dominated by the ethos of serviceable disposability, the implicit metaphysics is inimical to the inherent worth of things.[12] This implicit metaphysics needs to be countered with an explicit one—not the kind of metaphysics rightly decried by Heideggerians, where being is logically dominated in a closed system, but Desmond's kind of

11. See Desmond's critique of reductive empiricism in *Being and the Between,* 299–330.

12. See Desmond's discussion of unavoidable metaphysics in *Being and the Between,* 3–46.

metaphysics in which being cannot be so dominated because we affirm its excessive richness.

I think of the science and nature cartoons available to children today. Many of them are charming and well-meaning. They are no substitute for direct experience of nature, however, even if that experience is just walking through a city park or planting a seed in a paper cup. They certainly do not cultivate contemplative patience, given their nonstop action and half-hour format. In my more pessimistic moods, I think of how they tap into childlike wonder, perhaps even cultivate it to some limited degree, but in the service of a paradigm that would be hard pressed to offer a coherent account of the inherent worth of things, including our own worth. Indeed, the dominant educational paradigm seems to take the inherent worth of being for granted on the one hand, while on the other hand undermining it through a mix of subjective relativism, pragmatic utilitarianism, and scientistic materialism.

Where is something metaphysically and experientially richer to be found? Where are the sympathetic cultivators of wonder? There are promising, albeit still isolated, initiatives in schools, communities, and youth organizations. There are of course many parents, grandparents, extended family, neighbors, and friends who try to cultivate children's sense of wonder. We should work for broader movements, but we should all strive, at the very least, to be one of these companioning adults. It is a rewarding role to play, since as we guide children we will be guided by them in turn. We can exchange with them the gift of wonder.

We need more than a non-reductive science, of course. For even such a science, as important as it is, should not be our only approach to being. As other contributors in this volume suggest, we need religion that cultivates wonder at the richness of the world and the mysteries of existence, that cultivates reverence and humility and contemplation and asceticism of voracious desire.

As a humanities professor, I would stress that we also need a renewed sense of the import of art and literature. Returning to my own life, it was literature that stirred my sense of wonder again, that helped reopen my porosity, late in college. I was especially moved by nature poetry and by novels of rural communities. These compelled me to take a second look at my own upbringing, to reconsider its goods and its shortcomings. When I arrived in graduate school, I was sometimes told that my conception of literature was overly mimetic. I never quite bought this dismissal, though, even as I acknowledged that we shouldn't simply measure art against an external referent. It seemed too obvious to me that literature does change how we see the world, others, ourselves. I had experienced this.

Desmond offers supple resources here as well. He holds that art does not simply depict the richness of being—it incarnates this richness. The artwork reawakens us to the "too-muchness" of being first and foremost by inviting us to contemplate its own richness and mystery. The artwork has its own inherent worth in this regard, its own power to strike us with wonder. The artwork, however, always communicates with the wider world, and it can therefore attune us to richness beyond the work.[13] An Emily Dickinson poem about a winter afternoon's changing light, for instance, can make me more attentive to the light outside my own window.

Desmond holds that poetry preserves the excess of being. This means that any philosophy, any science, any theology that wants to preserve this excess will need a poetic register. He situates his own philosophy "between system and poetics." I think as well of the many poetic passages in Carson's *The Sense of Wonder*. And I recall my daughter's poetic description of the "tornado storm of bugs." Children need adults that will hear their poetry and share poetic words with them in return. Such poetry is another crucial way to cultivate wonder.

Bibliography

Carson, Rachel. *The Sense of Wonder*. New York: Harper & Row, 1965.

Desmond, William. *Being and the Between*. Albany, NY: SUNY Press, 1995.

———. "The Dearth of Astonishment: On Curiosity, Scientism, and Thinking as Negativity." In *The Voiding of Being: The Doing and Undoing of Metaphysics in Modernity*, 96–125. Washington, DC: The Catholic University of America Press, 2020.

———. *Ethics and the Between*. Albany, NY: SUNY Press, 2001.

———. *The Intimate Universal: The Hidden Porosity among Religion, Art, Philosophy, and Politics*. New York: Columbia University Press, 2016.

———. *Perplexity and Ultimacy: Metaphysical Thoughts from the Middle*. Albany, NY: SUNY Press, 1995.

Knepper, Steven. *Wonder Strikes: Approaching Aesthetics and Literature with William Desmond*. Albany, NY: SUNY Press, 2022.

Romanyshyn, Alexandra. "Metaxology and Environmental Ethics: On the Ethical Response to the Aesthetics of Nature as Other in the Between." In *William Desmond's Philosophy between Metaphysics, Religion, Ethics, and Aesthetics: Thinking Metaxologically*, edited by Dennis Vanden Auweele, 303–15. Cham, Switzerland: Palgrave Macmillan, 2018.

13. See the discussion of art and mimesis in Desmond, *The Intimate Universal*, 60–115. There is not room here to discuss the range and nuances of Desmond's account of art. I attempt this in my study *Wonder Strikes*.

4

The Astonishment of Philosophy

WILLIAM DESMOND
AND ISABELLE STENGERS

Simone Kotva

BY VIRTUE OF A remarkable coincidence, two recent critiques of scientism—by William Desmond and Isabelle Stengers—have at their center
the concept of counterfeit wonder. No companionable nor collegial solidarity might explain this coincidence, for these are thinkers of markedly different expression—stylistic and philosophical but also spiritual (and, one
suspects, political). If there is companionship here, it is thus less as regards
commitments to shared values and more as concerns faithfulness to what
Claude Romano[1] (following A. N. Whitehead, Gilles Deleuze, Alain Badiou,
and many others) calls an event: in this case (the case of Western science),
the ecstasy of limitless wonder reformed to justify the limitless pursuit of
knowledge at any cost. In lieu of hidden alliances, then, it is a companionship whose meaning emerges in the contiguous affects that constitute it:
the crossings of (Catholic) self-abandonment to the "too-muchness"[2] of being with (mystico-pragmatic) attention to the irreducible messiness of the
world as it is. Such, at least, is the intuition guiding this essay.

---

1. Romano, *Event and World.*
2. Desmond, *The Voiding of Being,* 119 [36].

61

Matrix of Science

Desmond's text, entitled, "The Dearth of Astonishment: On Curiosity, Scientism, and Thinking as Negativity," appeared in his 2020 volume, *The Voiding of Being*. What marks this text, which is meant to be taken as a critique of nihilistic thinking in Western philosophy and culture, is an attempt to resist the determination of wonder. By "determination," Desmond means any attempt to explain or pin down wonder, making wonder subsidiary to a higher cause: paradigmatically, in the West, the cause of the scientific enterprise. "Scientism" is what happens when wonder becomes construed as the handmaiden to science, rather than as its matrix, a construal that captures (and captivates) wonder—even as scientism appears to celebrate wonder by praising the virtues of "curiosity." Presented as "curiosity," Desmond argues, scientism "contracts the meaning of wonder," reducing wonder to investigative procedures that yield not "a kind of secret love of the object or subject of curiosity, but . . . a kind of hatred of its resistant mystery or marvel."[3] The "horror" and attendant "perplexity" at the world felt by canonical Western critics of this tendency (such as Kierkegaard, Nietzsche, and Dostoevsky) are thus to be understood as reactions to curiosity, not as instances of wonder.[4] Or rather, both curiosity and perplexity can be seen as modalities of wonder, overlapping yet distinct from astonishment, the third modality of wonder with which Desmond invites us to think, and which he sees congruent with the insights of mystical theology:

> It is not first that we desire to understand. Rather we are awoken or become awake in a not-yet-determinate minding that is not full with itself but filled with openness to what is beyond itself. . . . This is what makes possible all our determinate relations to determinate beings and processes.[5]

This concept of astonishment as that which facilitates rather than as a (merely) epistemic faculty[6] can be seen as further elaboration of the "metaxology" that Desmond lays out in his trilogy of books, *Being and the Between*, *Ethics and the Between*, and *God and the Between*. Like the *metaxu* ("in-between")—Plato's elusive name for the desire (*eros*) enabling emergent realities—astonishment escapes determination,[7] being itself the condition for every determinate meaning: astonishment is what makes relation

3. Desmond, *The Voiding of Being*, 120 [37].

4. Desmond, *The Voiding of Being*, 114 [32].

5. Desmond, *The Voiding of Being*, 109 [27].

6. Desmond, *The Voiding of Being*, 107 [25].

7. Plato, *Symposium* 203e.

possible. But what is at issue in this text is also, if not first of all, a practice, a "metaphysical mindfulness," as Desmond puts it; it is something like a way of living or being, necessitating a general rethinking of the relationship between epistemology and ontology that touches on the nature of scientific method. With epistemology no longer at an abstract remove from the terrain of being and beings, this mindfulness roots science in the indeterminateness of life: "*beyond expectation*"[8] there is "a more ontological astonishment that seeds metaphysical mindfulness."[9] But insofar as this wonder is essentially cultivated, it cannot be reflected accurately by phenomenology: "at issue is . . . a mindful *passio essendi* prior to and presupposed by every [endeavor] of the mind desiring to understand this or that."[10]

What is the nature of a method defined not by the object of wondering ("I'm interested in *this* or *that* . . .") but only by wonderment, its cultivation and affects? Simone Weil, the philosopher and mystic whose own writings—both on the *metaxu* and on scientism—rehearse themselves obliquely in Desmond's work, suggested that while science may be said to have instrumentalized wonder, metaphysics had done little better by elevating wonder to the extraordinary, leaving aside the whole question of human beings as sustained by wonder as a mere adjunct to this claim.[11] It is clear that what is at stake in Desmond's work is not simply to pit metaphysics against science, but rather to discover a zone of mindful practice that would alter the whole shape of the theology-science debate.[12] And it is precisely this question, this issue of mindful practice, that Stengers suggests conditions both science and philosophy, and which her essay on the failings of scientism defines as "the appreciation of this, always this, concrete situation accompanied by the halo of what may become possible."[13]

Relearning the Art of Hesitation

Stengers' book, which will be my sole subject for the rest of this essay, bears the title *Another Science Is Possible: Manifesto for a Slow Science*, first published in French in 2013. Unlike Desmond's piece, it is not concerned with

8. Desmond, *The Voiding of Being*, 107 [25].

9. Desmond, *The Voiding of Being*, 105 [24].

10. Desmond, *The Voiding of Being*, 109 [27].

11. Weil, *On Science, Necessity and the Love of God*, 65–70. On the *metaxu*, see for instance Weil, *Gravity and Grace*, 145–47.

12. On a zone of science–theology practice, see my article in the companion volume to this collection: Kotva, "Science and Theology in the Field."

13. Stengers, *Another Science*, 130.

metaphysics, nor does it seek the remedy for science in the discourse of philosophy (whether religious or phenomenological), and yet the tone is un-
apologetic: another science *is* possible ("une autre science est possible!").[14]
The key concept here is "possibility," by which Stengers (who, before turning
to philosophy, trained in the lab) intends something quite different to an
imagined, "improved" scientific community, or, for that matter, to any of
the existing "alternatives" to Western medicine (fated to follow the logic of
the very behemoth they defy—Stengers, by her own avowal, has "no time
for the New Age");[15] instead, "possibility" indicates a way of practicing that
would be open to the logic of possibility itself. Really open; that is, ready to
be thrown completely by the new, to slow down, hesitate, even to err from
the straight and narrow path set down by the dictates of scientific "rigor."
The ability to hesitate is perhaps the principal operative metaphor with
which Stengers defines her theory of possibility.

Hesitation is almost entirely lacking in the rhetoric of science, which
prefers to symbolize the scientist's motivating curiosity instead by the imag-
ery of the *vocation* ("vocation" is Stengers's word, throughout her book, for
the modality of wonder Desmond calls "curiosity," while Stengers' "curiosity" may refer *both* to the capture of wonder by scientism *and* to the ontologically constitutive desire Desmond names "astonishment"). It has been
noted that much of contemporary science operates on a secularized model
of monastic vocation: the laboratory researcher who is told to sacrifice all
for the "golden egg" (Stengers' term) of truth is rehearsing a model according to which hesitation is a sin. (Carolyn Merchant and Elizabeth Potter
have noted how Francis Bacon, Isaac Newton, and the other Puritan scientists during the English Civil War viewed hesitation as the sign of weakness but also of incontinence; the aim was not only to achieve greatness—to
"conquer" and "lay bare" the secrets of nature—but to resist temptations by
remaining modest and obedient to the higher values directing the work.)[16]
It is less well known just how such modelling works, today, to direct interest
in ways that block the very ability that science purports to celebrate when
it eulogizes—as its founding moment no less—a child's ability to marvel at
the world. Stengers even suggests that what we have in the vocation-speak

14. Citing from the French title: *Une autre science est possible! Manifeste pour un ralentissement des sciences.*

15. CRASSH Cambridge, "CRASSH: Magic and Ecology Podcast with Isabelle Stengers," YouTube video, 1:31:54. December 16, 2020. https://www.youtube.com/watch?v=N8IALHq_Kcs&t=648s.

16. Merchant, *The Death of Nature*; Potter, "Modelling the Gender Politics in Sci-ence"; Potter, *Gender and Boyle's Law of Gases*. Potter is discussed by Stengers, *Another Science*, 33–34.

of the sciences is a process that amounts to grooming. In *Another Science Is Possible*, after developing her critique of vocation, Stengers writes:

> Thinking of the way in which scientific institutions try to en-
> courage a taste for the sciences, one could almost speak—dare
> I say—of a kind of paedophilia, a thirst to capture the soul of
> the child. It associates science with a taste for strange gadgetry
> and disinterested questions, with the thirst for understanding
> and for science as a big adventure. Such tastes are, of course, no
> longer on the agenda by the time students enter university, and
> even less so when they start thinking about a research career.
> Far from being treated as a primary resource that is now under
> threat, young researchers of either gender, doctoral students or
> postdocs, have to accept the realities of onerous working con-
> ditions and fierce competition. They are supposed to grin and
> bear it: *the great adventure of human curiosity presented to them
> as children is replaced by the theme of a vocation that demands
> body-and-soul commitment.* And this is what we accuse today's
> young people of no longer accepting: compliance with the sacri-
> fices that service to science demands.[17]

Scientific institutions capitalize on the "soul of the child," redirecting an initially indecisive openness into a decisive commitment (vocation) to ideals that only scientific institutions can fulfil: "a taste for strange gad-getry and disinterested questions," and so on. As a result, the child who was once open to the absolute, able to hesitate and be swept away by the "great adventure of human curiosity," is now open absolutely only to the ide-als of science: the child's curiosity, initially receptive to whatever life threw at them, is now able to receive only that which scientific institutions tell them they ought to be interested in. If we recall the distinctions between the different modalities of "wonder" in Desmond's typology, we can recognize this as the moment astonishment becomes curiosity, and wonder becomes captive to determination. But there is also Desmond's modality of wonder-as-perplexity, discernible in the way Stengers describes the adult researcher who faces the disillusioning realities of "onerous working conditions and fierce competition." Much of what Stengers has to say in *Another Science Is Possible* concerns the fatal effects of this disillusionment on the young researcher attempting to "make it" in science. Will they manage to put aside their childish dreams and quell any lingering doubts about "the point of

---

17. Stengers, *Another Science*, 24–25. My emphasis.

it all"? Or will the doubts get the better of them, shunting the would-be scientist into "softer" disciplines, into the humanities or social sciences?[18]

Being Able to Notice: Slowing Down

In *Another Science Is Possible* Stengers generally defines the quality of hesitation in terms of its effects on the speed of thought: a person who hesitates is by definition unable to charge ahead at full speed—they are taking a break, looking around, "[paying] attention to what may lurk."[19] They are slower than their peers, less efficient and less productive. In the direct pipeline that exists between scientific institutions and industry in West-led global capitalism, this is of course a capital point, and where the matter at hand—the matter of revoking regimes of counterfeit curiosity—takes on political and ecological, as well as philosophical and hermeneutic, urgency. In the capture of curiosity by the scientific institution, curiosity is refigured in ways that are structurally analogous to the repurposing of environments, societies, artefacts, and organisms by capitalist economies. Rather than perceived as gift, curiosity is seen as resource and treated accordingly, directed at this or that aim, and, most importantly, made subservient to criteria of speed and efficiency.

Stengers argues that metaphors such as "discovery," "conquest," and "exploration," when applied to science as well as to intellectual labor more broadly (one thinks of the overuse of "exploration" in academic articles), reinforce this ideal and increase institutional resistance to alternative ways of practicing descriptive of curiosity rather than of vocation. In *Another Science Is Possible,* after developing the age-old image of the scientist interested in the things themselves, she writes: "If relevance rather than authority or objectivity had been the name of the game, the sciences would have meant adventure, not conquest."[20] If one keeps in mind the simultaneously focussed and meandering nature of adventure—the way adventure lacks a specific goal and yet is characterized by a passionate love for specifics (the adventurer, unlike the conqueror, does not have the eye on the prize and is easily distracted by the places they pass through)—we can see that the childish dream, "the great adventure of human curiosity," carries out a precise political intention.

18. Stengers, *Another Science*, 23–47.

19. Savransky and Stengers, "Relearning the Art of Paying Attention," 136.

20. Stengers, *Another Science*, 144.

Staying with the Trouble of Curiosity

We can now see the particular urgency of Stengers' essay. The scientist's supreme gesture is to consign curiosity to a childhood fantasy, that is, to consider curiosity as in some manner unreliable and in need of authorial guidance (science as avuncular figure leading a puerile researcher by the hand). But what does it mean for unreliability instead to appear as method? And in what sense does Stengers' essay express this method?

Stengers begins by detailing what we could have pictured, namely, that to insist on unreliability is to enforce "irony, perplexity, and guilt," encouraging "the retreat to purely academic and inconsequential postmodern games" among those who "no longer endorse this conquering and missionary enterprise."[21] To the contrary, unreliability designates precisely the refusal to settle for such disavowal: "they [who have given up the ideas that have blessed this enterprise of modern science] also have to reclaim a different, positive, definition of themselves and of civilisation, in order to regain relevance and become capable of weaving relations with different people and natures."[22] At this point, Stengers gives a counter-factual story imagining a different science by means of making reference to the need for every scientist to have a "dream" of a better future. Immediately afterward, as if realizing the insufficiency of this programmatic statement and fearing that her words might seem excessively utopian, she takes recourse to the work of Donna Haraway, the feminist historian of science:

> What I call [my dream] is not the solution to this difficulty, but a name for it, a name calling the invention of modes of gathering that complicate . . . by introducing hesitation. This is what Donna Haraway has now turned into a thought-provoking motto: "staying with the trouble."[23]

"Staying with the trouble" indeed is the motto of Donna Haraway's recent work, *Staying with the Trouble: Making Kin in the Chthulucene*, in which Haraway thinks with—among many others—the remarkable ethologist Vinciane Despret. A brief reading of these pages suffices to realize what attracts Stengers to Haraway's motto where hesitation is concerned, and how it may be of importance to us here. First of all, Haraway-Despret distinguishes clearly between curiosity, the ideal captured by scientific institutions, and the virtue of openness: "the energetic work of holding open the possibility that surprises are in store, that something *interesting* is about to

21. Stengers, *Another Science*, 141.
22. Stengers, *Another Science*, 141.
23. Stengers, *Another Science*, 151.

happen, but only if one cultivates the virtue of letting those one visits intra-actively shape what occurs."[24] Despret's work on animal behavior is remark-able for questioning the assumptions that go into the practice of observing scientists. Instead of assuming that the observing scientists should focus their attention on this or that animal, Despret argues that they should, to the contrary, enlarge their attention so as to enrich the field of possibilities in order to let the animal, along with the scientist (and all other possible co-agents), shape the interaction. Tongue-in-cheek, Haraway calls Despret's "a curious practice":[25]

> Despret's cultivation of politeness is a curious practice. She trains her whole being, not just her imagination . . . "to go visit-ing." Visiting is not an easy practice; it demands the ability to find others actively interesting . . . to cultivate the wild virtue of curiosity, to retune one's ability to sense and respond—and to do all this politely. [Creatures studies by science] are not who/what we expected to visit, and we are not who/what were anticipated either. . . . Asking questions comes to mean both asking what another finds intriguing and also how learning to engage *that* chances everybody in unforeseeable ways.[26]

What makes Haraway-Despret remarakble is precisely this state of curiosity, which cannot be attributed to preconceived notions of what a scientist ought to be curious about. It is significant that Haraway refers to this way of practicing science as "staying with the trouble," using an expres-sion that implies complexity and mess and that indicates the implication of science in the very objects it studies: "[This] kind of thinking enlarges, even invents, the competencies of all the players, including [of the scien-tist], such that the domain of ways of being and knowing dilates, expands, adds both ontological and epistemological possibilities, proposes and enacts what was not there before."[27] This is why Stengers can speak of curiosity as a practice that is speculative even as it is concrete, since "this is not so much a question of manners, but of epistemology and ontology, and of method

24. Haraway, *Staying with the Trouble,* 127. It should be pointed out that Haraway's essay on Despret was first published in 2015: thus, Stengers—whose essay came out in 2013—is not referring to *Staying with the Trouble* but to discussions preceding it. However, in order to weave the two together more closely it seemed appropriate to refer to *Staying with the Trouble* in this instance. For Despret's work on the practice of observing scientists, see Despret, "The Becoming of Subjectivity in Animal Worlds," 123–39. Stengers also refers to Despret in *Another Science,* 71.

25. Haraway, *Staying with the Trouble,* 127.

26. Haraway, *Staying with the Trouble,* 127.

27. Haraway, *Staying with the Trouble,* 126–27.

alert to off-the-beaten-path practices."[28] And it is in relation to Haraway's work that Stengers' elusive reference to an "other" science in the title of her book becomes fully comprehensible. Starting with the feminist historians of science and primatologists that influence Haraway's early research,[29] Haraway's work is motivated by the untiring and exuberant attempt to narrate science—in close dialogue with the minor and marginalized sciences and scientists of our time—according to "ordinary stories, ordinary becoming 'involved in each other's lives."'[30] Haraway calls this practice "attunement" and defines it as a simple organic capacity of affectability, that, like the capacity of surprise, "must be made without guarantees or the expectation of harmony with those who are not oneself—and not safely other, either."[31]

It goes without saying that such a definition of science cannot remain inattentive to "faith." The fact is that the practice described by Stengers (and Haraway-Despret) seems to come to light only in moments that are aesthetic or spiritual, and that are made possible by a kind of unconditional trust. But even the next example, which is meant to show the proximity between this way of practicing and ritual, bears on a special case, one that lies in the vicinity of action rather than belief. "The Quakers did not quake before their God," writes Stengers,

> but before the danger of silencing the experience that would disclose what was being asked of them in a particular situation, before the danger of answering that situation in terms of predetermined beliefs and convictions. The crucial point . . . is not, it seems to me, the belief in some supernatural inspiration we might feel free to snigger at. The point is the efficacy of the ritual, an aesthetic one, enhancing what Whitehead called "the concrete appreciation of the individual facts in their full interplay of emergent values"; or the appreciation of this, always this, concrete situation accompanied by the halo of what may become possible.[32]

Perhaps one might say that the difficulty of staying with the trouble of curiosity by means of surprise leads us instead into an area that is even more surprising, in which both science and religion present us with the elusive practice of attentiveness as such. As Stengers puts it elsewhere, "it

28. Haraway, *Staying with the Trouble*, 127.

29. Haraway, "Situated Knowledges," 575–99; Haraway, *Simians, Cyborgs and Women*.

30. Haraway, *Staying with the Trouble*, 76.

31. Haraway, *Staying with the Trouble*, 76.

32. Stengers, *Another Science*, 129–30.

sounds very mystical, although it is really very practical (like every mystic, in fact)."[33]

Neither Authority nor Faith

It is now possible to say a few more words about the sense in which we were able to remark, at the beginning of this essay, that there is in the thought of both Desmond and Stengers a shared commitment rather than a sharing of ideas, and that this sharing—its form and expression—has strong bearing on the subject of our enquiry. To begin with, it will be necessary to read Desmond's critique of scientism neither as a philosophical nor as a theological response to scientism but rather in light of Stengers' call to a mode of activity beyond authority and faith. I do not think that reading in this way diminishes the value of either thinker, for the truth of each is put at risk only insofar as it is taken to be an exhaustive representation of a reality exceeding determination. Need I remind that Desmond's philosophy, the philosophy of astonishment *par excellence*, "knows" this, professes it painstakingly on page after page? Yet a philosophy of astonishment is not the same as the astonishment of philosophy, and often it is neither desire nor hospitality but the professing of surprise that frustrates the irruption of the unforeseen. I wonder if it is not in reading together in this way that the conditions for surprise described by both texts can best be facilitated: for Desmond, a "metaphysical mindfulness" preceding every judgement, and for Stengers, a practice where what matters is neither agreement nor consensus but the ability to think together? Reading in this way I found myself needing now to discern not similarities nor differences but degrees of openness: the openness, to one another, of thinkers considered as creature-kin thinking together, even in cases where such openness is made obscure by the disciplines through which thought inevitably is formed.

This is the promise but also the problem of a philosophy of astonishment, the ambiguity contained in the revocation of counterfeit wonder and the reclaiming of curious practices across disciplines, cultures, and creatures.

Bibliography

CRASSH Cambridge. "CRASSH: Magic and Ecology Podcast with Isabelle Stengers." YouTube video, 1:31:54. December 16, 2020. https://www.youtube.com/watch?v=N8IALHq_Kcs&t=648s.

33. Pignarre and Stengers, *Capitalist Sorcery*, 142.

KOTVA THE ASTONISHMENT OF PHILOSOPHY 71

Desmond, William. *Being and the Between*. Albany, NY: State University of New York Press, 1995.

———. *Ethics and the Between*. Albany, NY: State University of New York Press, 2001.

———. *God and the Between*. Oxford: Blackwell, 2008.

———. *The Voiding of Being: The Doing and Undoing of Metaphysics in Modernity*. Washington, DC: The Catholic University of America Press, 2020.

Despret, Vinciane. "The Becoming of Subjectivity in Animal Worlds." *Subjectivity* 23 (2008) 123–39.

Haraway, Donna. *Staying with the Trouble: Making Kin in the Chthulucene*. Durham, NC: Duke University Press, 2016.

———. *Simians, Cyborgs and Women: The Reinvention of Nature*. London: Routledge, 1991.

———. "Situated Knowledges: The Science Question in Feminism and the Privilege of Partial Perspective." *Feminist Studies* 13.1 (1988) 575–99.

Kotva, Simone. "Science and Theology in the Field." In *New Directions in Theology and Science: Beyond Dialogue*, edited by Peter Harrison and Paul Tyson, 36–55. London: Routledge, 2022.

Merchant, Carolyn. *The Death of Nature: Women, Ecology and the Scientific Revolution*. San Francisco: HarperCollins, 1979.

Pignarre, Philippe, and Isabelle Stengers. *Capitalst Sorcery: Breaking the Spell*. Translated by Andrew Goffey. Basingstoke, UK: Palgrave Macmillan, 2011.

Potter, Elizabeth. *Gender and Boyle's Law of Gases*. Bloomington, IN: Indiana University Press, 2001.

———. "Modelling the Gender Politics in Science." *Hypatia* 3.1 (1988) 19–33.

Romano, Claude Romano. *Event and World*. Translated by Shane Mackinlay. New York: Fordham University Press, 2009.

Savransky, Martin, and Isabelle Stengers. "Relearning the Art of Paying Attention: A Conversation." *SubStance* 47.1 (2018):130–45.

Stengers, Isabelle. *Another Science Is Possible: Manifesto for a Slow Science*. Translated by Stephen Muecke. Cambridge: Polity, 2018.

Weil, Simone. *Gravity and Grace*. Edited and translated by Emma Crawford and Mario von der Ruhr. London: Routledge, 2006.

———. "Scientism: A Review." In *On Science, Necessity and the Love of God*, edited and translated by Richard Rees, 65–70. London: Oxford University Press, 1968.

5

Astonishment and the Social Sciences

Paul Tyson

WILLIAM DESMOND'S ANALYSIS OF curiosity and astonishment, and his appreciation of the manner in which our experience is always porous to transcendence and our knowledge is never finally a mastery, is a vital component of the way I pursue research in the social sciences. This makes me an outlier in the social sciences, for reasons we shall touch on, but a Desmondean approach is very much compatible with a deep interest in the intrinsically fascinating human worlds of day-to-day life. In fact, I think it is only when our approach to the human worlds of meaning, action, and power are both open to what modern empirical knowledge and modern interpretive theory can *reveal* and aware of what such knowledge and interpretation also *conceals* that the social sciences best serve the pursuit of truth. In what follows I shall first outline the normal parameters of the social sciences as presently practiced, and then sketch what I think a Desmondean improvement on that normality looks like. I will conclude with a short description of how I have found scholarship as a Desmondean sociologist to be very intellectually satisfying, at the same time as being necessarily marginal to the central norms of the social sciences that now dominate our academies.

There are—in broad terms—two trajectories in the contemporary social sciences, one of which is positivist, the other of which is best described by the German term *vestehende soziologie* (interpretive sociology). The positivist stream is indebted to French theorists like Comte and Saint-Simon, and also to the pragmatic, materialist, and analytical trajectories in recent Anglo-American thinking. Positivists are very good at the measurement and statistical analysis of correlations between social conditions and social behaviors and attitudes. The empiricist social scientist usually treats these correlations as functions of physical necessity, revealing law-of-nature-like "social mechanics" that describe the workings of "society" at an aggregated trend-trajectory level. In contrast, the *vestehen* (understanding) approach to the study of socially situated meanings and practices is indebted to German theorist like Dilthey, Weber, and Husserl. Here the interpretation of socially situated meanings and actions cannot be reduced to the measurable categories of determinate causation, such that any social science that is reductively positivistic performs a serious category error. Interpretive sociologists suspect that positivists invalidly reduce cultural meanings and practices to sets of interactive material facts. But human society is more complex than a chemical reaction (complex though that be, of course). *Vestehende* sociologists suspect that positivists naively expect that an objective statistical analysis of the physically measurable causal dynamics of power and functionality in specific human societies tells you all you need to know about what society is and how it works. To the contrary, interpretive sociologists maintain that there is something intrinsically indeterminate, meaning concerned, and contextually specific about different human societies. Society is not a machine.

In practice, positivist and interpretive sociological trajectories often have quite a bit to do with each other. This, of course, is a good thing. Taking something of an objective stance and looking at and measuring what is actually occurring in any given sociological context allows for a more factual and determinate analysis than trying to simply "understand" what is going on in virtue of being embedded in the meaning structures of any given sociological form-of-life. On the other hand, a pure objectivity is not actually obtainable when it comes to time, location, community, and practice-embedded human meanings. Hence, paying close attention to the contextually defined meanings of facts (and to the meaning of what constitutes a significant human fact and what does not, in any given context) will make your data-set more intelligible, will define its interpretive limits more carefully, and will hopefully produce some caution for policy makers and social actors who may naively wish to treat society like a machine that can be epistemically mastered, engineered, and then simply "driven" towards certain desired ends. Humility should follow from an appreciation that facts and interpretations, whilst not reducible to each other, never stand in full autonomy from each other in the social world, or in the construction of social scientific knowledge.

The move towards a more conscious synthesis of positivist and interpretive approaches to sociology is what some theorist describes as "reflexive sociology."[1] Here, a sociologist will conduct detailed quantitative and statistical analysis of social phenomena, and yet such sociology will be reflexive in that it is seeking to be consciously aware of its own interpretive commitments when it produces sociological knowledge, and when it constructs and interprets sociological data-sets.

A Desmondean sociology, however, cannot be characterized as either positivist or interpretive, nor as a reflexive and metaphysically agnostic synthesis of the two. This is because Desmond not only has a richly dialectical understanding of the relationship between facts and interpretations, but he also has a metaphysically open understanding of reality. It is this metaphysical openness that is glaringly absent from both the positivist and the interpretive traditions expressed in most contemporary sociology scholarship.

Being born in the nineteenth century, our contemporary social sciences appear after Kant, within the Western intellectual tradition. This is highly significant. Kant situates the empirical *a posteriori* world of facts, the *a priori* world of logical necessity, and the practical and action-concerned interpretive worlds of human meanings, all within the domain of phenomena (the world *as it appears to us*). That is, Kant renders reality itself (as independent of our knowing and acting) unknowable. This is a decisive turning point in modern philosophy, the post-metaphysical turn.

After Kant, transcendence no longer speaks to us (or so the dominant categories of secular modernity seem to believe).[2] That is, we seem to have learnt how to interpret all transcendence as "transcendental," where the structures of phenomena are defined by the workings of *our* mind and awareness, not by reality itself. Our conscious awareness has become a closed epistemic system such that sensory facts are always facts as they appear to me. Likewise, rational, moral, metaphysical, and theological truths are always logical necessities, values, speculative beliefs, and ultimate meaning horizons as situated within *my* contextually situated and interpreting mind. In the shadow of Kant, both the positivist and the interpretive traditions of sociology are closed to transcendence. The empirical world is now a non-realist construction of our knowledge such that science itself can no longer be a tentative (perhaps analogical) account of how things really are, and where any correspondence between perception and reality must start and finish with *our* theory-embedded perception, and whatever utility *we* can get out of it. The world of human meaning is now *essentially*[3]

---

1. Josephson-Storm, *The Myth of Disenchantment*, 11.

2. On the manner in which immanence is unavoidably embedded in transcendence in our actual experience of the world, see Tyson, *Seven Brief Lessons on Magic*.

3. I use the word "essentially" here ironically, and with considerable grief. That

contextually, pragmatically, and semiotically defined, with no metaphysical remainder. Our academy now largely accepts that an anthropocentric and finally autistic enclosure defines the limits of human meaning in any positivist and interpretive context. Now, verbal and textual signs *and* what those signs signify are locked entirely within humanly constructed systems of meaning, which—due to the finally solipsistic interpretive atomism of each knower—always entails unavoidable semantic slippage between the speaker and the hearer, and never finally locates meaning within any partial participation in real quantitative, qualitative, or interpretive truth. In this context, Derrida's understanding that words are always located in the playful domain of symbolic construction, with no finally fixed meanings, and Žižek's understanding that no single line of meaning-construction ever fully integrates with any other line of meaning-construction, such that any idea of global meaning on which language itself is premised is an oppressive political construct, are both perfectly sensible Kantian stances.[4] But, it is a pity we can't take either Derrida's or Žižek's stances as *true* (if they are right). For truth itself is now obsolete. Perhaps we should wonder if Kant is right?

It is the exclusion of transcendence from sensible reality and human meanings that renders both positivist and interpretive sociology as "closed" epistemic systems. These closed systems, as sophisticated and complex as they are, remain starved of truth in the transcendental semantic wilderness of the epistemically mastered and solipsistic phenomenological enclosure of each knower's own mind. But Desmond shows us another way.

---

is, after Kant the very meaning of "essence" in the Western philosophical tradition is made to mean its opposite. In traditional Platonist/Aristotelian metaphysics, essence concerns intellective and qualitative forms that are non-contingent, metaphysically transcendent, and only ever partially expressed in space and time. Form is expressed, partially, in matter, and poetically known by us (via sensation and revelation) to some degree, but form/essence itself is exactly not contextually, pragmatically, and culturally constructed. By making phenomena the only thing we can know, the material world as it appears to us becomes the *essential* truth of things, rendering all meaning and all knowing contextual, transient, and constructed. That is, we now only know a world *without* metaphysical essence, and this absence of transcendent essence (ironically) is to us the "real" (essential) meaning of the world we perceive. Thus the meaning of reality and essence in their pre-Kantian philosophical traditions, come to mean their opposites from the turn of the nineteenth century on. This is a great irrationality, a great irrealism. For how can the essential meaning of reality be that it has no essential meaning? And how can any "realism" be entirely dependent on—and trapped within—our perspective? Essence and reality—as words that attach to no transcendent point of reference—are now meaningless words. Postmodern irrealism and the impossibility of textual and linguistic meaning—and hence the reduction of language to power—follow from this high modern innovation as night follows dusk.

4. Derrida, *Of Grammatology*; Žižek, *The Ticklish Subject*.

Kant's shift from transcendence to the transcendental is an intellectual sleight of hand. He gains epistemic closure (mastery over the domain of our own phenomenological "world") at the cost of abandoning openness to unmasterable Reality reaching into our small phenomenological worlds, and touching us with astonishment and revelations that we cannot master or fully say. For astonishment and inspiration are signature traces of the impact of the Real on our minds. That is, defined and containable knowledge and meaning within the phenonomenological realm is bought—after Kant—at the cost of shutting out transcendent reality and preventing any sort of Real Truth from seeping into our small worlds of human meaning-construction. Curiosity is the desire to containably know what things are and how they work. The temptation of curiosity towards epistemic sin occurs when the desire to know and control becomes a lust for totalizing determinate epistemic and use mastery within defined domain boundaries. When we give way to this lust, curiosity overcomes wonder. Kant makes the domain of human knowledge masterable by limiting it to phenomena, but he offers us no compelling reason why we should believe that wonder about the Real itself has become obsolete.

Hamann gave us excellent reasons not to believe Kant at the outset, and Desmond demolishes Kant's post-metaphysical turn in his book *The Intimate Strangeness of Being*.[5] Yet, to Desmond, just because Kant's rejection of the partial knowability of metaphysical truth is spurious, this does not mean that curiosity and artificially limited epistemic mastery must be excluded from knowledge. In fact, curiosity and epistemic and semantic construction are invaluable partners with Reality in all human knowing and meaning. But the order must be correct; curiosity is the servant of, and subservient to, astonishment. For it is wonder and astonishment that produces the desire for knowledge, not knowledge that produces wonder and astonishment. That is, the Real reaching into our frame of knowledge is what gives our knowledge its partial participation in true meaning and valid reasoning, it is not the (always limited) reach of our knowledge into the Real that determines the value and meaning of human reason and knowledge.

The social sciences, after Kant, are metaphysically disabled. Here, we limit knowledge to what we think of as the epistemically masterable domains of modern positivist empiricism, cultural constructed meanings, and

---

5. Johann Georg Hamann demolished Kant's *Critique of Pure Reason* back in 1784 with his "Metacritique on the Purism of Reason" (see Hamann, *Writings on Philosophy and Language*, 205–18) on linguistic grounds. The reasons for Kant's "success" as a turning point in modern philosophy have always had more to do with its appeal than its actual merit. For some pointed critical analysis of the supposed post-metaphysical turn from Kant onward, see Desmond, *The Intimate Strangeness of Being*, 89–119.

mathematical analytic models, even though we ironically assume that all these domains are phenomenologically insular (and hence, artificial constructs). But, as Desmond points out, our hubris prefers the small kingdom of masterable curiosity and constructed meanings to the boundless domain of wonder. For we can preside, as a god, over our own knowledge- and meaning-constructions, defined within the contained parameters that we set, whereas we are forever a child (of God) in the vast—sometimes frightening—ocean of wonder. We *want* our knowledge constructions to be closed systems over which we preside, and that give us control over our world; we thus readily close ourselves off to the obvious reality of transcendence and our obvious participation in unmasterable Reality when we are constructing firmly mastered knowledge.

Desmond points out that the most basic feature of our experience of being a conscious and communicative being is astonishment. The immanent is porous to the transcendent, time flows over eternity, the material is embedded in the spiritual, the knower is in analogical communion with intelligible reality, myth is more hermeneutically basic than fact, the I is always in communion with the divine Thou; these are the truths that we naturally respond to with astonishment and gratitude. Astonishment defines us as worship beings; beings who find the center and source of worth as beyond themselves. Because this is how we actually experience the world, social sciences that function in reductively positive and objectivist categories (which, as subjects, we are not capable of experiencing) or phenomenologically contained subjective meaning categories (as if transcendence did not saturate our actual phenomenological landscape) miss the most defining characteristics of the actual reality of the social and human worlds. But this is not an innocent missing of the mark. It is idolatry (self worship) which is the defining epistemic sin of the fallen human condition.[6] If we can just imagine the human worlds of meaning and value as if there is no source of truth, meaning, and value above our own minds, the original epistemic temptation to unreality (we shall be as God) is—and oh so easily—repeated.

Let us briefly look a bit further at why the modern social sciences are committed to the Kantian epistemological rejection of metaphysics.

John Milbank makes the point that social theory in the Western tradition is already theological, and that the Kantian post-metaphysical turn is only possible because of the (unsatisfactory) theology of the Enlightenment's secular modernity. To see the world as if faith, divinity, and the miraculous do not exist is to display a range of metaphysical and theological

6. For a most helpful exploration of epistemic sin, see Kierkegaard, *The Concluding Unscientific Post-script*.

commitments that define the interpretive framework of post-Kantian knowledge. These commitments don't just arise from nowhere, they are bequeathed to Western modernity from late medieval Western theology.[7] And—speaking as a Christian theologian—Milbank argues that the unsatisfactory theological origins of secular modernity are heretical, and can only be properly corrected by orthodox Christian theology. The salient point here, for my argument, is that the commitment to human knowledge and human interpretation being non- or anti-theological, and non- or anti-metaphysical, is not theologically or philosophically neutral, but is *an inescapably theological and metaphysical commitment*. Yet so committed to "secular reason" has our academy become that it has become "heretical" for us to do overtly theologically grounded social theory any more, even though all social theory in the Western intellectual tradition actually *is* theologically premised!

There is, of course, an obvious hubristic advantage to doing social theory with anti-theological and anti-metaphysical theological and metaphysical commitments. We can pretend that we are at the same time above human knowledge and meaning, as its source, and that there is no meaning or truth above our own designations. This we think of as objective and philosophically neutral, and we impose these social scientific objective and neutral epistemic language conventions onto all players who would enter the domain of social theory and social scientific knowledge discourse. The ideas that one must (a) have theological and metaphysical commitments in order to actually do social theory and that (b) such commitments give us a *better* picture of our actual experience of the human world than the fiction of non-committed objectivity and constructivist hermeneutic relativism could give us, are excluded by the language games of the post-metaphysical and secular social sciences. Our small world of artificially limited social theory (limited to our own knowledge and meaning constructions) is the triumph of determinate and defined curiosity over apophatic and en-wondered astonishment.

Leaving the secular and post-metaphysical social sciences to play their own language games, let us now ask, what would sociological data and a good interpretive understanding of the human categories of thought, meaning, and value look like were we to have a Desmondean sociology?

A Eucharistic[8] (literally, "good gift") Christian sociology starts from the awareness that there is a Giver who gifts us to reality, and reality to us.

7. Milbank, *Theology and Social Theory*.

8. My thanks to Sotiris Mitralexis for enabling me to think of knowledge in Eucharistic terms.

To a Eucharistic sociology, we are called into being by the only One who *can* call us into being, and who also gives Himself to us as the life of the world.[9] An awareness of ontological gratitude correlates with the primary insight that our very personhood is a function of being gratuitously spoken to, formed by, and relationally bonded to a Value, Purpose, and Intelligible Meaning (i.e., the person of the Divine *Logos*) that is beyond both human knowing and beyond a merely material knowledge of the cosmos itself. If we take ontic gratuity as the first reality of our existence, then the theological and metaphysical *grounds* of our and nature's being informs our mind from above, at the same time that observable facts and the contingently situated meanings of social reality informs our mind from below. But this is an important point; we do not prove higher meaning and purpose with lower scientific facts and sociological meanings (as if the locus of truth and reality is—in the Kantian fashion—*my* perceiving and rational mind alone). Rather, lower facts, values, and meanings become *more* intelligible when we lay ourselves open to receive an analogical vision of the kind of high metaphysical and theological reality that does, in existential actuality, gift us with our innate astonishment towards being itself.

All the great natural theologians of the Christian tradition start—as Thomas Aquinas does—with God and move from there to nature. This is the only reasonable direction of truth seeking inferential movement in such matters. That modern natural theology, which attempts to prove God from nature and rational human knowledge, always fails is no accident, failure is an inherent feature of this fundamentally misguided enterprise.

So let us presuppose—as a matter of prior interpretive and doxological commitment—that the grounds of being is transcendently good, intelligent, and loving, and that our most primary personal reality concerns the doxological relation of each individual to the transcendent source of their own self-hood, which in turn, frames the way we love (or fail to love) our neighbor. This is how Kierkegaard does sociology in modern times, and—as one would expect of a Lutheran—this is embedded in an Augustinian approach to the human world.[10] If we start from this position we are starting from honest, overt, and orthodox theological and metaphysical commitments from which we can then look at the facts of sociological data and the interpretation of contingent sociological meanings, values, and purposes. These last two terms are interesting because we got rid of values and

9. Schmemann, *For the Life of the World*.

10. Kierkegaard, *The Sickness unto Death*; Kierkegaard, *The Two Ages*; Tyson, *Kierkegaard's Theological Sociology*; Ross, *Gifts Glittering and Poisoned*; Bell, *The Economy of Desire*.

purposes as features of nature (they are now artefactual and entirely contingent constructs of culture) when we got rid of Aristotle in the seventeenth
century. But actually, value and purpose have never gone away from reality
just because our new instrumental approach to factual knowledge (now
called "science") can't see them.

If we start from theologically and metaphysically rich enough starting
points, then we can construct reasonably demonstrable factual knowledge
theories of the real material world, and poetically enact and interpret the
partially grasped values and meanings that facilitate true human flourishing. Granted, these are no longer categories of thought compatible with our
"scientific" view of the social sciences, but this is because we have an impoverished understanding of science that is "merely" positivistic, and hence,
is fully discrete from wisdom. Further, wisdom itself has been reduced to
cultural meaning glosses that have no objectivity and no bivalency, but are
really (so our materialist theories assume) functions of physical necessity
and imagination, and are not partially grasped real values and true meanings about the proper end of a good human community, as was once believed. We now try and do social science without any genuinely qualitative
categories of the common good, the highest good, or goodness at all, as
a moral reality.[11] The outcome of social scientific knowledge without any
genuinely qualitative moral truths is social scientific "knowledge" that is
blind to wisdom, and that facilitates either a purely instrumental manipula-
tive understanding of sociological power, or that spirals off into an endless
relativism of meaning interpretations that renders all interpretations as
ultimately equally meaningless. So our social scientific "knowledge" is not
a knowledge of qualitative and purposive human reality—the reality we are
actually existentially embedded in—but it is "knowledge" of abstract sociological models, which happen to lack any meaningful moral and purposive
categories. But is such knowledge a knowledge of human reality at all?

As regards reforming the existing social sciences in a Desmondean
direction, a few brief concluding comments can be made. There is now an
enormous body of sociological data collected by various government and
private agencies in an attempt to manage and administer the modern nation state effectively, and by business and interest groups (including governments) seeking to harness the energies of mass consumer society in the age
of surveillance capitalism for commercial gain and cultural and political influence.[12] This data facilitates a technologizing of society, which, as Jacques

11. See Rist, *Real Ethics*, for a very helpful enterprise in recovering moral realism
in our times.

12. Zuboff, *The Age of Surveillance Capitalism*.

Ellul pointed out, has the dangerous problem that we now readily become the objects and servants of our own tools of knowledge and power.[13] But there remains a scholarly helpful side to all this data. If I want to know how many people in Australia are living in single-occupancy households (about a quarter of Australian households, by the way); if I want to know what our real unemployment dynamics are like (this is actually *not* disclosed by our highly tailored official government statistics, but some digging around reveals that only about than 8 percent of the Australian economy comes from agriculture and manufacturing, so what sort of work are we all meant to do?); if I want to theorize about despair as a feature of consumer society and want statistics on self-harming and suicidal behavior as a useful correlate to my argument; if I want to know what has happened to income distribution and house prices over the past four decades, then the data is readily available. So, from a data perspective, the social sciences have never had more fuel in their information tank from which to run their knowledge-construction engine. But now we have the problem that we can't see the wood for the trees. For even with super-computer processing, the sort of data processing Google does is only good for a simple instrumental end—profit—however sophisticated the preference- and attention-monitoring (and shaping) algorithms are. If wisdom does not guide knowledge, knowledge becomes meaningless, and dangerous, and can from there easily reduce to sub-human and finally demonic power.

Data handling and statistical awareness are important skills for the modern social scientist, and the Desmondean social scientist needs to be good at these skills. However, no matter how good one is, the Desmondean social scientist will have another problem. In seeking to do social science starting from a theological and metaphysical openness to transcendence, I find it easy to be dismissed as an outlier who has no proper reverence for the disciplinary boundaries of modern knowledge-construction. If I was a proper theologian, I would only concern myself with doctrine, scriptures, and religiously situated ethics. Theologians should—so both our seminaries and our secular academies often wrongly assume[14]—stay within the domain of religion and not venture out into economics, politics, metaphysics, epistemology, aesthetics, science, and technology, etc. If I was a proper philosopher, I would play by the rules of pluralism that makes no final claims to ultimacy (largely assuming Kant's post-metaphysical stance) but

13. Ellul, *The Technological Society*; Ellul, *Propaganda*.

14. "Religion" is—as Peter Harrison puts it—not a natural kind, hence its domain cannot be defined in any sort of scientific or essentialist manner (the same is true for "science"). See Harrison, *The Territories of Science and Religion*, 4.

that constructs tightly reasoned arguments moving from generally defend-
able secular and scientific starting points to a defended conclusion, that can
stand among other positions in the market place of secular and scientific
reason-constructs (without any claim to ultimacy). Philosophers can't do
philosophy if they overtly start philosophizing from committed religious
or metaphysical first principles. If I was a proper positivist social scientist,
I would only make careful observation-dependent and statistically valid
models of trends and attitudes in our society, without passing any normative judgement on what those models meant. I would not analyze society
from moral realist and overtly metaphysically committed starting points. If
I was a proper interpretive social scientists, I would treat meaning-making
as entirely defined by culture (in the Kantian post-metaphysical manner)
and would not seek to find traces of transcendence reaching into human
culture and revitalizing social constructivism.

The insistence on siloed and artificially defined knowledge discourses
in which scholars may be recognized by their similarly siloed peers as experts is the natural corollary of the absence of metaphysics and theology
from knowledge itself. For some ultimate meaning and reality discourse is
necessary to unify knowledge itself, and without such a unification, fragmentation is not only inevitable, it is mandatory.[15] For this reason, thinking about society from metaphysically theological commitment premises
cannot be overtly performed (even if it cannot be covertly avoided) in the
post-metaphysical and secular academy.[16] So when, as a Desmondean soci-

15. For a fascinating exploration of what might be called the anti-metaphysical
tyranny of modern tolerance, see Conyers, *The Long Truce.*

16. Interdisciplinary scholarly enterprises are interesting endeavours in our
contemporary academic contexts. If we get theologians, philosophers, and scientists
together—as the "After Science and Religion" project that I coordinated from the University of Queensland did—the combination of first-order global disciplines (theology
and philosophy) with specialist knowledge disciplines (such as biology, chemistry, and
physics) can be wonderfully fruitful, but only if there is a tacit functional unity among
the participants regarding first-order commitments. Interdisciplinary scholarship is
unworkable if there is no first-order common ground. If, say, Buddhist, Christian,
and atheist theologians and utilitarian, Augustinian, and Kantian philosophers were
pluralistically distributed among both the first-order and specialist-knowledge participants, no meaningful interdisciplinary outcome could arise. Pluralism is always
only relatively possible within some overarching grounds of commonality. Pluralism is
not incommensuratism. Only systems of knowledge and belief that are to some degree
commensurate can live together in a community of plural interpretations. Incommensuratism is a very radical monism where each one is forever alone in their knowledge
and meaning construct. Post-metaphysical and secular fragmentation does not permit of dialogical difference in harmony (i.e., genuine pluralism), but only permits of
multiple incommensurate discourses, each one of which is a gnostic and linguistically
defined monologue.

ologist, I reason from theological and metaphysical commitments grounded in a faith tradition of astonishment and openness to the divine fount of
wonder, this violates the rules of the academy in the specialist cataphatic
categories of the discipline's post-metaphysical norms, and it is offensive to
modern knowledge-construction in virtue of my scholarship assuming that
there really *is* a unity to all true knowledge and meaning. For, as Dorothy
Sayers pointed out, any coherent knowledge system requires a synthesizing
vision of unity that orders all knowledge, which also provides a meaningful
formational purpose to knowledge itself.[17] This does not mean that wonder
is ever contained within any single frame of theological or philosophical
doctrines and ideas, and it does not mean that different schools of first-
order metaphysical and theological commitment cannot wonderfully learn
from each other, but Confucius is simply correct to notice that one cannot
build common meanings, common knowings, and common actions with
people of a different Way.[18]

So the Desmondean social scientist, embedded in the first-commit-
ments of metaphysical and theological wonder, and overtly conducting
scholarship as a doxological enterprise, must remain an outlier to the modern secular discourses of the social sciences. Yet, our prevailing (idolatrous)
epistemic priesthood very much needs troubling prophets in the wilderness. The future, in one way or another, lies with truth (or at least, falsehood

17. Sayers, *The Lost Tools of Learning*, 19. In this beautiful short description of
what Americans now call a Classical education, Sayers points out that in the West's
intellectual tradition "Theology is the Mistress-science, without which the whole
educational structure will necessarily lack its final synthesis." Theology provides the
different aspects of a good education with its unifying telos. Without such a synthesizing telos, the entire enterprise of education has no point. The model of education that
prevailed in Britain, even into the early twentieth century, understood this feature of
education, as that model arose from the monastic missions that produced Christen-
dom and the universities of the Western intellectual tradition. It is a concerted late
nineteenth-century secularism, and a reductively immanent naturalism acting as a
new and secularized unifying telos (an ersatz theology), that finally displaces genuine
theology from its centring and unifying role in the purpose and meaning of education. On this defining feature of the late nineteenth-century re-configuring of the very
meaning and purpose of education, see Chadwick, *The Secularization of the European
Mind in the 19th Century*.

18. Confucius, *The Analects of Confucius*: Analect 15.40 ". . . with those who follow
a different Way, to exchange views is pointless." That is, fundamentally different understandings of the Way of Heaven (the truth of how things really are) do not permit
a unity of knowledge, value, meaning, or purpose. Though, functional unities may still
be achieved without this essential unity. Analect 9.30 explains: "There are people with
whom you may share information, but not share the Way. There are people with whom
you may share the Way, but not share a commitment. There are people with whom you
may share a commitment but not share counsel."

is the cause of its own demise), and so if the prophets speak some measure
of truth, they will prevail, even though they are outliers to their own times.
In this manner, William Desmond's vision of astonishment is a vital contribution to a better approach to the social sciences for the future. And I am
delighted to say that, as far as I can see, a new approach to the social sciences
is being born in our times as theologians recover their metaphysical and
global nerve, and refuse to be contained within the cataphatic and discrete
intellectual silo of "religion."[19]

Bibliography

Bell, Daniel, Jr. *The Economy of Desire*. Grand Rapids: Baker, 2012.
Cavanaugh, William. *Being Consumed*. Grand Rapids: Eerdmans, 2008.
Chadwick, Owen. *The Secularization of the European Mind in the 19th Century*.
    Cambridge: Cambridge University Press, 1975.
Confucius. *The Analects of Confucius*. Translated by Simon Leys. New York: Norton,
    1997.
Conyers, A. J. *The Long Truce*. Waco, TX: Baylor University Press, 2009.
Derrida, Jacques. *Of Grammatology*. Baltimore: John Hopkins University Press, 2016.
Desmond, William. *The Intimate Strangeness of Being*. Washington, DC: Catholic
    University of America Press, 2012.
Ellul, Jacques. *Propaganda*. New York: Vintage, 1973.
———. *The Technological Society*. New York: Vintage, 1965.
Goodchild, Philip. *Theology of Money*. Durham, NC: Duke University Press, 2009.
Hamann, J. G. *Writings on Philosophy and Language*. Cambridge: Cambridge University
    Press, 2007.
Harrison, Peter. *The Territories of Science and Religion*. Chicago: University of Chicago
    Press, 2015.
Josephson-Storm, Jason. *The Myth of Disenchantment*. Chicago: University of Chicago
    Press, 2016.
Kierkegaard, Søren. *The Concluding Unscientific Post-script to the Philosophical
    Fragments*. Princeton: Princeton University Press, 1992.
———. *The Sickness unto Death*. Princeton: Princeton University Press, 1980.
———. *Two Ages*. Princeton: Princeton University Press, 1978.

19. That English language theologians are escaping from the death trap of the
'religion' box and venturing into politics, economic, etc—as theological metaphysicians—has a lot to do with John Milbank's startling and still fresh intervention into
social theory back in the early 1990s. See Milbank, *Theology and Social Theory*. A few
excellent examples of sociological interventions from the grounds of Christian metaphysics are: Goodchild, *Theology of Money*; Ross, *Gifts Glittering and Poisoned*; Bell,
The Economy of Desire; Cavanaugh, *Being Consumed*. Of course, thinkers not writing
in English have been less effected by the late 19th century reduction of knowledge to
science than the Anglophone world—see for example Jacques Ellul and Paul Ricoeur.
Johannes Hoff, writing in German, shows this sort of living trajectory at the time of
writing this book.

Milbank, John. *Theology and Social Theory*. Oxford: Blackwell, 2006.
Rist, John M. *Real Ethics*. Cambridge: Cambridge University Press, 2002.
Ross, Channon. *Gifts Glittering and Poisoned*. Veritas. Eugene, OR: Cascade, 2014.
Sayers, Dorothy L. *The Lost Tools of Learning*. 1948. Reprint, Louisville, KY: GLH, 2017.
Schmemann, Alexander. *For the Life of the World*. New York: Saint Vladimir's Seminary Press, 1973.
Tyson, Paul. *Kierkegaard's Theological Sociology*. Eugene, OR: Cascade, 2019.
———. *Seven Brief Lessons on Magic*. Eugene, OR: Cascade, 2019.
Žižek, Slovoj. *The Ticklish Subject: The Absent Centre of Political Ontology*. London: Verso, 1999.
Zuboff, Shoshana. *The Age of Surveillance Capitalism*. London: Profile, 2019.

6

Curiosity, Perplexity, and Astonishment in the Natural Sciences

Andrew Davison

WILLIAM DESMOND'S REMARKABLE ESSAY "The Dearth of Astonishment: On Curiosity, Scientism, and Thinking as Negativity"[1] asks its reader how she is going to respond to creation—to the gift of being as such—and what it asks of her, and awakens in her. Like so much of Desmond's writing, the essay offers a form of spiritual exercise to any reader willing to approach it that way, here focused on what he calls three "modalities of wonder": curiosity, perplexity, and astonishment. With a larger project of criticism in view, the essay also addresses the stifling effects of scientism, which result in a "dearth of astonishment." This rings true, although I do not have the expertise to comment on such a large scale myself. Rather, my purpose in this essay is to consider one culture or field of enquiry I do know well, namely the natural sciences, and to suggest that Desmond's sense of the full breadth of wonder applies to that world, including his perplexity and astonishment, and not only the curiosity that he finds troubling. I do so by noting the close alignment between Desmond's analysis and the still-illuminating account of the nature of science in Thomas Kuhn's *The Structure of Scientific Revolutions*.[2] Desmond's category of curiosity mirrors Kuhn's "normal science," his perplexity marks out periods when an era of normal science comes up against its limits, and his astonishment is characteristic of Kuhn's "paradigm shift,"

1. Desmond, *The Voiding of Being*, chapter 3, 96–125 [15–42].
2. Kuhn, *Scientific Revolutions*.

by which we both see the world as very different from how we previously understood it, and find a new beauty and elegance in how a new perspective can integrate the previously explicable with the previously inexplicable.

Let us attend further to what Desmond identifies as three ways in which the human being can be struck by wonder: as curiosity, perplexity, and astonishment. Curiosity views the world as intellectually safe, orderly, and open to gentle probing, or puzzling. When confidence in our capacities and traditions of interpretation proves unequal to the task, we move into perplexity, finding what lies around us intractable to our attempts to order it (as Desmond sees it, rather by a surfeit of meaning than a shortage).[3] Finally, with astonishment, that surfeit serves to awaken us to a wonder now in its highest modality, within which we set aside a desire to confine and control in favor of a posture of awed reception.

Such a description falls far short of Desmond's charged, poetic expression. He writes, for instance, that "Curiosity tends to be oriented to what is determinate or determinable, perplexity to something more indeterminate, astonishment to what is overdeterminate, as exceeding univocal determination,"[4] or of "a more primal ontological astonishment that seeds metaphysical mindfulness," that in "restless" perplexity "thinking seeks to transcend initial indeterminacy toward more and more determinate outcomes," and that in curiosity "the initiating openness of wonder is dispelled in a determinate solution to a determinate problem."[5]

Desmond is careful to distinguish between scientism and the practice of the natural sciences themselves. By scientism, he means an attitude, fundamentally philosophical in nature, that supposes the natural sciences to be able, in theory, to answer any question.[6] That tends both to limit which questions are deemed reasonable to set, with those that are not open to the scientific method being discounted, and to an inevitable reductionism and dismissal of the metaphysical. Desmond points to two related examples: to evolutionary psychology, with its fallacy that if evolution gave us minds then every thought (and everything thinkable) turns out to be about evolution, and to "cybernetic scientism," by which "the mind is to be likened to a computer" or, indeed, is said simply to be one, which can be taken so far as denying that thought is meaningful, since it is only data.[7]

3. Desmond, *The Voiding of Being*, 108 [26].

4. Desmond, *The Voiding of Being*, 96–97 [16].

5. Desmond, *The Voiding of Being*, 105 [24].

6. Desmond, *The Voiding of Being*, 97 [16].

7. Desmond, *The Voiding of Being*, 98 [17].

Desmond's principal charge against scientism is that it sets aside questions about being, is resistant to mystery or marvel, and that it tends to make things in the world "serve our instrumental desire, and hence be serviceable, but when used, they are used up, and are hence disposable."[8] In this, the natural science themselves seem not to be the culprit. "Scientism is not science but an interpretation of science," Desmond writes.[9] Yet, an association between scientism and science is never entirely set aside, and Desmond opens his essay by aligning the practice of the sciences with curiosity, rather than only their philosophical or cultural interpretation: "What looks like an insatiable curiosity seems to drive the scientific enterprise. In principle no question seems barred to that curiosity."[10] I do not doubt the stifling effects of scientism, reductionism, and a "dearth of astonishment," but I want in this chapter to strengthen the separation of scientism from the practice of the sciences, not least because my persistent experience has been that professional scientists are by no means always scientistic and that, when they are, that is shaped more by their place in polite, educated, middle-class society than it is by their work as scientists.

Even after sixty years, Thomas Kuhn's *Structure of Scientific Revolutions* strikes me as getting to the heart of the work of natural science, whether in periods of sturdy development, crisis, or revolutionary change. By showing that Kuhn's analysis would bring the work of scientists under the whole Desmond's analysis—that this human work and these human reactions can be characterized by all three modalities of wonder—I hope to underline that the natural sciences as such are not an exercise that stultifies metaphysics or the human capacity for astonishment.[11]

According to Kuhn, most science is undertaken within the parameters of "normal science."[12] This is the usual state of scientific activity, when underlying assumptions—about how to think and model the world, how to design experiments and equipment, and how to go about interpreting one's results—are under no particular challenge. Scientific labors in such a period

8. Desmond, *The Voiding of Being*, 120, 123 [37, 39].

9. Desmond, *The Voiding of Being*, 97 [16].

10. Desmond, *The Voiding of Being*, 96 [15].

11. To that a reply might be made that the natural sciences are still concerned with beings, and their properties, not with being-as-such. There is truth to that, but it strikes me that a delight and astonishment before being in some degree is not unconnected with, or unproductive of, delight and astonishment before being-as-such, with its metaphysical register. Kuhn notes that it has been in periods of crisis through to paradigm shifts (of perplexity turning to astonishment) that scientists have been most open to philosophy, with its wider metaphysical interests (*Scientific Revolutions*, 88).

12. Kuhn, *Scientific Revolutions*, 5.

bring about small, incremental elaborations. Anomalous results may be turned up, but they will tend to be rationalized, or set aside.[13] If the number of such anomalies mounts, however, running even to observations that cannot be explained at all under prevailing assumptions, a scientific culture will eventually find itself in a period of crisis. In due course, a dam may break, leading to a shift in thinking that is no longer comparable to the incremental development in understanding previously described, constituting instead a fundamental revision of underlying concepts: what Kuhn calls a paradigm shift. Acceptance of that shift will take time, and meet resistance, but in science one set of paradigms generally wins out against another, ushering in another period of normal science, conducted confidently under the new paradigm

Kuhn offers several historical examples to justify this analysis, including the emergence of Copernican astronomy, Lavoisier's oxygen theory of combustion, and relativity.[14] I would point to the triumph of Newtonian physics up to the end of the nineteenth century, followed by an accumulation of observations that could not be explained in those terms, and the paradigm shift to quantum mechanics (alongside relativity). The period before the crisis was one of "normal science," since, however great the advances (and there were many creative extensions of the Newtonian paradigm into new areas), underlying assumptions about mass, energy, and so on, remained the same. This was a period of confidence, with a sense that the whole world could be described in terms of bouncing balls and vibrating springs. By the late nineteenth century, however, anomalies were beginning to accumulate, for instance the "ultraviolet catastrophe" (according to which, on Newtonian expectations, an object at thermodynamic equilibrium ought to radiate infinite amounts of energy as light), the photoelectric effect (the observation that electrons are not ejected from a metal by light, however intense, while it remains below a certain frequency, characteristic of each metal, and that the energy of the electrons ejected at higher frequencies depends on the frequency of the light, but not its intensity), and the temperature dependence of specific heat capacities (the capacity of a substance to absorb heat given a certain change in temperature is found to drop unexpectedly at low temperatures).[15]

13. Kuhn, *Scientific Revolutions*, 78.

14. Kuhn, *Scientific Revolutions*, 68–69, 69–72, 72–74.

15. Kuhn notes Albert Einstein's comment that "before he had any substitute for classical mechanics, he could see the interrelation" between anomalies in all three of these areas (*Scientific Revolutions*, 89).

As Kuhn suggests would usually happen in a period still of normal science, at first the string of anomalies tended to be discounted. Eventually, however, a group of scientists dared to ask a more radical "what if?," proposing that various properties of matter, previously thought to be continuously variable, such as energy, might be found only in a ladder-like range of discrete values. (They are "quantized.") Approached in those terms, those previously inexplicable phenomena turn out to be exactly what one would expect.[16]

Much of what marks out curiosity for Desmond, for better or worse, is characteristic of Kuhn's normal science. For instance, it is built on self-assurance, being "predicated on the assumption that the scientific community knows what the world is like."[17] An association with the solving of puzzles offers another, particularly strong alignment with curiosity: "Perhaps the most striking feature of the normal research problems [of normal science] . . . is how little they aim to produce major novelties, conceptual or phenomenal."[18] The regnant paradigms of normal science are "a criterion for choosing problems that, while the paradigm is taken for granted, can be assumed to have solutions. To a great extent these are the only problems that the community will admit as scientific or encourage its members to undertake."[19] If normal science is generally "eminently successful in its aim, the steady extension of the scope and precision of scientific knowledge,"[20] that is largely because it is so careful about the questions it asks, bringing us close to the "dearth" of Desmond's title.

Despite what we might think at first, however, Desmond is not simply and only dismissive of curiosity, going so far to write that it plays an "indispensable" part in cognition.[21] Although curiosity can slip over into something deadening or idolatrous, "it is necessary to assert the *constitutive* role of curiosity in getting as univocal a grip as possible on the intelligibility of being, in addition to as precise an articulation as possible of that

16. Of the ultraviolet catastrophe, the radiation of light is capped because, beyond a certain point, the energy levels corresponding to any higher frequency of emission will not be occupied; of the photoelectric effect, only once light is of a frequency with sufficient energy can electrons be freed from their resting energy state in relation to the attraction of the nucleus; of specific heat capacity, certain modes of storing energy (such as vibration in a gas) are shut off below a certain temperature, once a transition between even the first and second rungs of the energy ladder is off limits.

17. Kuhn, *Scientific Revolutions*, 5.

18. Kuhn, *Scientific Revolutions*, 35, reiterated in the later postscript (205).

19. Kuhn, *Scientific Revolutions*, 37.

20. Kuhn, *Scientific Revolutions*, 52.

21. Desmond, *The Voiding of Being*, 116 [33].

intelligibility."[22] There can be a "good sense" of curiosity, which "finds things interesting and surprising; its desire to know is open to the novel and the strange. It fastens on things in their interesting determinacy. It is open to what is unfamiliar and odd."[23] Consider this in relation to Kuhn's description of normal science, which is not to be despised either, as "the activity in which most scientists inevitably spend almost all their time,"[24] and one that has done much to furnish improvements in many aspects of daily life.

Just as Kuhn's description of normal science looks very much like Desmond's vision of curiosity, so his description of science in a time of crisis is characterized by something close to Desmond's idea of perplexity.

> Sometimes a normal problem, one that ought to be solvable by known rules and procedures, resists the reiterated onslaught of the ablest members of the group within whose competence it falls. On other occasions a piece of equipment designed and constructed for the purpose of normal research fails to perform in the anticipated manner, revealing an anomaly that cannot, despite repeated effort, be aligned with professional expectation. In these and other ways, normal science repeatedly goes astray.[25]

The tendency, at first, to paste over these challenges demonstrates something of the relative lack of imagination in the endeavor of normal science, and its fear of risk and the inexplicable, all redolent of Desmond's "curiosity."[26] Eventually, if observations of anomalies pile up, these rogue phenomena may be accepted as genuine, but that would only highlight further the presence of what Desmond calls a "gap," across which we find ourselves unable to travel, or even an "ontological horror" in view of that which is in "excess to our rational measure."[27] These periods of crisis and anomaly are typically ones of "pronounced professional insecurity," given that the puzzles of normal science turn out not "to come out as they should."[28]

22. Desmond, essay as previously published, 116, emphasis in original.

23. Desmond, essay as previously published, 117.

24. Kuhn, *Scientific Revolutions*, 5.

25. Kuhn, *Scientific Revolutions*, 5–6.

26. "Much of the success of the enterprise [of normal science] derives from the community's willingness to defend that assumption, if necessary at considerable cost. Normal science . . . often suppresses fundamental novelties because they are necessarily subversive of its basic commitments" (Kuhn, *Scientific Revolutions*, 5). On a propensity to make ad hoc adjustments, at first, rather than revise one's paradigms, see 78.

27. Desmond, *The Voiding of Being*, 114 [31]. On reactions to X-Rays, for instance, see Kuhn, *Scientific Revolutions*, 59.

28. Kuhn, *Scientific Revolutions*, 67–68.

This brings us to Kuhn's paradigm shifts, and their association with astonishment. "When . . . the profession can no longer evade anomalies that subvert the existing tradition of scientific practice—then begins the extraordinary investigations that lead the profession at last to a new set of commitments, a new basis for the practice of science."[29] Paradigm shifts come about

> not by deliberation and interpretation, but by a relatively sudden and unstructured event like the gesalt [sic] switch. Scientists often speak of the "scales falling from the eyes" or of the "lightning flash" that "inundates" a previously obscure puzzle.[30]

Place that alongside Desmond's description of his preference for the word "astonishment," as one able to capture "the ontological bite of otherness. There is *the stress of the emphatic beyond expectation. . . .* There is something of the blow of unpremeditated otherness in being struck by astonishment."[31] A paradigm shift also resembles astonishment in the way it restructures thought itself—not simply some particular thought or thoughts—and leads the scientist to see the whole in new ways.[32]

Caution is needed, however, in alignment of paradigm shifts with Desmond's astonishment. The relation is more complex than the previous discussions of a relation between perplexity and crisis, or curiosity and normal science. For one thing, Kuhn's uneasiness with realism, in a metaphysical or epistemological sense, stops him from presenting a paradigm shift as the result of the world revealing itself to us anew, or as any sort of call upon us by an objective reality.[33] That stands in marked contrast with what, for me, is among the most appealing aspects of what Desmond offers in his essay, in his discussion of astonishment as reception and openness to the truth of things:

> We do not possess a capacity for wonder; rather we are capacitated by wonder. . . . I behold the majestic tree and exclaim "This is astonishing!" I am not projecting my feeling; rather, the tree is coming to wakefulness in me, while I am being awakened by the tree, and I am awakening to myself, in a more primal porosity.[34]

29. Kuhn, *Scientific Revolutions*, 6.

30. Kuhn, *Scientific Revolutions*, 122.

31. Desmond, *The Voiding of Being*, 107 [25], emphasis in original.

32. Kuhn, *Scientific Revolutions*, 111–35.

33. Kuhn, *Scientific Revolutions*, 205–7.

34. Desmond, *The Voiding of Being*, 107, 109 [25, 27].

To explore this, he picks up Plato's description of wonder (*thaumaze-in*) as the pathos of the philosopher: "a patience, a primal receptivity, . . . a pathos more primal than activity, a patience before any self-activity."[35] There is a real gap between Kuhn and Desmond here, although one could perfectly well recast Kuhn's account in realist terms (probably in "critically realist," or when I have called "mediated realist" terms),[36] even if he himself would not.

That suggests one way in which the theologian or philosopher of Desmond's stripe (or mine) might develop upon the openings offered by Kuhn's writing. In other respects, it may be Kuhn who offers suggestions for a further exploration of astonishment. If so, such creative provocation would underline the fruitfulness of considering Kuhn's vision of science as reflecting the full breadth of Desmond's vision of wonder, rather than as an example of curiosity alone. The first suggestion is that paradigm shifts are not only intellectual, in some highly conceptual sense, but also entail changes in ways of life.[37] Much could be said (and I doubt that Desmond himself would be slow to offer it) about the communal and social dimensions of astonishment and its consequence. A second, and related, point is that paradigm shifts cannot be explained solely in terms of deliberations about concepts, data, or interpretations. They also have what Kuhn describes as a personal or aesthetic dimension involving the character of particular people, and relationships with them.[38] What role, we might ask, following that lead, does the mediation of persons, for instance, have in opening us up to astonishment? Third, Kuhn offers an account of what it might take to talk across the gap between rival paradigms, which he approaches as an exercise in "translation."[39] Related to the previous point, then, what role can the already astonished person play in communicating that to others? Finally, Kuhn offers an appealing sense of the value of crisis in science. As he puts it, "one of the reasons why prior crisis proves so important [is that] [s]cientists who have not experienced it will seldom renounce the hard evidence of problem-solving to follow what may easily prove and will widely be

---

35. Desmond, *The Voiding of Being*, 109 [27].

36. Davison, "Science and Religion."

37. Kuhn, *Scientific Revolutions*, 94, 157, although see the caveat on this in the postscript (176–81). Kuhn likely has in mind changes to their form of life *in the practice of their science*. In the ancient world, and indeed a good deal later, the practice of what we would today call natural science was seen as at least potentially transformative of the person more widely, rather than separating off a sense of the person simply "*qua* scientist*" (Ip, "Physics as Spiritual Exercise").

38. Kuhn, *Scientific Revolutions*, 156.

39. Kuhn, *Scientific Revolutions*, 198–204.

regarded as a will-o'-the-wisp."[40] That might augment the otherwise rather dour role played by perplexity in Desmond's vision.

Scientism, then, is the dearth-dealing force, not the natural sciences as such. There is much in them of perplexity and astonishment, as well as curiosity.

Bibliography

Davison, Andrew Paul. "Science and Religion." In *The Encyclopedia of Philosophy of Religion*, edited by Charles Taliaferro and Stewart Goetz. 1st ed. Wiley, 2021.

Desmond, William. *The Voiding of Being: The Doing and Undoing of Metaphysics in Modernity*. Washington, DC: The Catholic University of America Press, 2020.

Ip, Pui Him. "Physics as Spiritual Exercise." In *After Science and Religion*, edited by Peter Harrison and John Milbank, 282–98. Cambridge: Cambridge University Press, 2022.

Kuhn, Thomas S. *The Structure of Scientific Revolutions*. 3rd ed. Chicago: University of Chicago Press, 1990.

40. Kuhn, *Scientific Revolutions*, 158.

7

Scientism as the Dearth
of the Nothing

Richard J. Colledge

WILLIAM DESMOND'S PAPER, "THE Dearth of Astonishment"[1] provides a welcome elaboration on a prominent strand of his thinking that he has been developing now for several decades. I refer here to his focus on various modes of mindfulness, or ways of being open to the ontological context within which human beings find themselves. These modes guide our basic ways of dwelling within the openness of being. They provide us with particular ways of "reconfiguring" or fundamentally orientating ourselves to the elemental "ethos of being," or what he calls the "between."[2]

Much of Desmond's paper is devoted to unpacking the metaphysical underpinnings of three such modes of attending to the between. Accordingly, human beings can attend directly and remain open to the overwhelming (since "overdetermined") richness of ontological givenness in a mode of *wonder*. As his many books articulate, such "metaxological mindfulness" is at the heart of authentic metaphysical, religious, ethical, and aesthetic ways of being and mind. At other times, there is instead a restless *perplexity* concerning the indeterminacy of the primal ethos of being, a restlessness born of a sense of being overwhelmed by overdeterminacy, the reflex to which is a desire to determine and domesticate the rich over-determined

1. Desmond, "The Dearth of Astonishment," which is chapter 3 in Desmond, *The Voiding of Being*, 96–125 [15–42].

2. See, for example, as early as Desmond, *Being and the Between*, ch. 1. For a more recent treatment, see Desmond, *The Intimate Strangeness of Being*, ch. 1.

whole. However, it is against the backdrop of a third modality of metaphysical mindfulness—*curiosity*, which, in Desmond's account, amounts to a double removal from primal wonder—that he locates the tacit ideology of "scientism."

Science and Scientism

Desmond's account of scientism is telling on various levels. In his interpretation, in this infinitely restless desire to conquer all equivocity, the *truly* infinite overdeterminedness of the given plenitude is effectively denied and is replaced by a "counterfeit" infinite of scientific inquiry and progress. But like all counterfeits, when the vital genuine becomes concealed behind the shimmering but impoverished artifice, the results can be hugely problematic.

At the core of Desmond's understanding of scientism is a claim about hubris. Scientism, in this construal, is not just a matter of curiosity that responds to the wonder of being. It is rather a matter of effectively denying the givenness of being as such, so that reality can be re-created in our own image. For to affirm such absolute givenness is to acknowledge that we dwell within (and because of) a givenness that vastly surpasses human contrivance. Instead, modern scientism insists on the absolute priority of human self-determination. There is thus, observes Desmond, "a secret form of tyrannical will to power behind the scientistic desire to reconfigure all of being according to its dictates of a certain globalizing univocity."[3] Yet this reconfigured ethos is entirely parasitic on the originary ethos of the between that is the whole context of our being in the first place. There is no true *causa sui*: all life is given to itself, and only on that basis is any reconfiguration possible. In rejecting the *humility* implicit in an open wonder at the givenness of being, the scientistic attitude interprets such fundamental reliance as a *humiliation*. This will not do. As Desmond puts it: "[t]he original wonder gifts us with the gift of being able to receive gifts. If it is contracted or mutilated in scientistic curiosity, then we cannot accept any gift, we cannot accept anything as given, for it is not given on our terms."[4]

The consequences of this complex are sketched briefly by Desmond in the final section of his paper. He links the energy of this scientistic hubris with the idol of *technē*: "the drive for technological sovereignty" qua "unhindered self-determination." Economics becomes untethered from the *oikos* of our granted earthly sustenance as it drives forward relentless technological imperatives, seeking to satisfy infinite consumption, assuming a

3. Desmond, *The Voiding of Being*, 124 [40].
4. Desmond, *The Voiding of Being*, 124 [40].

dogmatism of "serviceable disposability," and spurred on (I would suggest) by a myth of infinite available natural resources. He goes on to point out the irony of a military-industrial complex that marries economics with the threat of death and destruction even as the pharmaceutical industry strives to deny the *pathos* of human finitude. It is this finitude that highlights, more than anything, the conceit of the denial of human creatureliness.

Of course, it is crucial not *simply* to conflate such a critique of scientism with a critique of the natural sciences simpliciter. True, it is not difficult to recall key moments in the history of Western philosophy of science where the seeds of full-blooded scientism seem to have been planted. One might, for example, point to the dawn of modern scientific empiricism in Bacon's notion of the experiment as a "bring[ing] force to bear on matter," to "vex and drive it [*vexet atque urgeat*] to extremities,"[5] a view famously reiterated by Kant's notion of the scientist being "instructed by nature not like a pupil, who has recited to him whatever the teacher wants to say, but like an appointed judge who compels witnesses to answer the questions he puts to them."[6] Desmond seems to have something like this sense in mind when (apropos Heidegger's reference to "the piety of thinking") he remarks that "question[s] can be posed in the modality of love of the true, or in the modality of aggression, even hatred. There are ways of questioning that lack reverence for the thing questioned. They are impious."[7]

But does such impiety characterize modern science as such? This need not be the case. Such understandings of the nature of natural scientific inquiry are far from definitive. Scientific research may indeed be conducted whilst retaining a genuine sense of wonder at the superabundant richness of the natural world as given, and such an approach distinguishes the work of many scientists the world over, of both a secular and a religious inclination. *Yet* the seductive logic of scientific and technological sovereignty—that treats the natural world as a neutral assemblage to be epistemically conquered to the end of practical and commercial exploitation—clearly continues to hold sway to a disconcerting degree. In that sense, there is indeed a chilling "impiety" characteristic of the dominant trajectory of modern scientific thought and practice.

5. Bacon, *The Works of Francis Bacon*, 726.
6. Kant, *Critique of Pure Reason*, B xiii–iv, 109.
7. Desmond, *The Voiding of Being*, 120 [37].

The Natural Sciences and the Existence Question

One issue in particular lies at the heart of the gap between metaphysical wonder and the natural sciences, and in this way opens the way toward scientism. This relates to the intrinsically immanentist character of the methodology of the empirical sciences. The philosophical foundations of empiricism (of which the modern sciences are a tightly controlled and procedurally driven application) are devoted to the dictum that knowledge of the natural world comes only through close observations of the measurable behavior of things (and systems of things) in nature. Science is therefore intrinsically devoted not so much to "nature" as such, but to *natural things* in terms of inductive inferences concerning their characteristics and relations.

Thus, strictly speaking (outside of the wider *humanity* of the scientist), genuine metaphysical perplexity has no place in the sciences. The scientist may be perplexed about how to interpret seemingly contradictory data, for example, but this is not the kind of perplexity to which Desmond is referring. Intimations of transcendence have no standing in the scientific method, which deals with observation rather than intimation. Further, the very concept of transcendence is precluded by the programmatic focus on *things* in their immanent relations. What this means is that metaphysical questions of existence (as opposed to observations about the *actuality* of this or that natural thing) are rightly beyond the reach of modern science. Contingency can only concern immanent causality, and not the sheer gratuity of being that points toward an agapeic origin that cannot be grasped (brought to heal) conceptually and mathematically.

However, *so long as* science is understood as but *one* "way of being and mind" among others,[8] this is not a problem. What the scientific mind does, it does well, and it need not look to extend its reach beyond its own competence. It is not the function of science to deal with the ultimate question of existence. In dealing with *what* is, and *how* it is, the empirical and mathematical sciences deal with things and not with the question of the ground by which there *are* things. Science becomes *scientism* only when other crucial openings of mind to the world are denied; when the scientific method is absolutized and when, as Desmond notes, there is an "underlying presupposition . . . that being is determinate, and in a manner that invites science to make it as univocally precise as possible."[9] But *recognizing* scientism as a problem presupposes that the *limits* of scientific understanding are recognized. In the context of a prevalent tone-deafness to other modes of insight, such recognition is perhaps increasingly rare.

8. See Desmond, *Philosophy and its Others.*

9. Desmond, *The Voiding of Being,* 101 [20].

To take one famous and influential example, one might consider the concluding lines of Hawking's *A Brief History of Time* where this master explorer of the mathematical basis of the *physis* wonders *perplexedly* about why "the universe [would] go to all the bother of existing?"; about what it is that "breathes fire into the equations and makes a universe for them to describe."[10] This is indeed a key question that distinguishes driving scientific curiosity from genuine metaphysical perplexity (if not wonder). Yet in that very moment, Hawking returns immediately to deterministic (mathematical and empirical) science for answers to a question that legitimately points *beyond* causal determinacy. He thus ends his book with a romantic vision of the ultimate triumph of science (the desired "theory of everything") that would understand the conditions that made the Big Bang possible, and that would thus purportedly end all need for metaphysical perplexity.

Such absolutization of scientific inquiry serves as a perfect illustration of Desmond's quite specific understanding of scientism. Hawking's dreamed-of "ultimate triumph of human reason" would mean that we would "know the mind of God."[11] It is no accident, of course, that this phrase invokes echoes of Hegel's logistic absolutism that similarly points to the ideal of absolute knowing. Yet (to utilize a key distinction Desmond has often made in his work), even if such a grand "theory of everything" was possible, it would still only be a profound insight into an extreme moment of the *becoming* of the universe. It would not be convertible with insight into the ultimate ground of the radical *coming-to-be* of anything at all. Ultimate "why" questions are not simply reducible to the most difficult "how" questions.

Existence and the Nothing:
On Wonder and Heideggerian Porosity

Of Desmond's various *modern* philosophical interlocutors in the paper, two figures in particular haunt his reflections. One, of course, is the ubiquitous G. W. F. Hegel. Desmond provides us here with a useful recapitulation of his already well-attested critique of Hegel's purported ("counterfeit") dialectical mediation of indeterminacy through self-becoming, here understood as itself a form of "speculative scientism."[12] However, in what follows, I will seek to draw out the particular relevance of the other recent figure who appears several times in Desmond's reflections. In doing so, I wish to build upon

10. Hawking, *A Brief History of Time*, 174.
11. Hawking, *A Brief History of Time*, 175.
12. Desmond, *The Voiding of Being*, 101 [20].

Desmond's own assessments of Martin Heidegger's openness to the notion of ontological overdetermination.

The radical question of existence is raised in the context of the "nothing." It is *that* there is not just nothing (i.e., utter absence or void) that is the point of wonder; all the more so when being is not just given, but given with such profuse richness. This is a point to which Desmond alludes in his reference to "primal astonishment before the marvel of the 'to be' as given—given with a fullness impossible to describe in the language of negativity, though indeed in a certain sense it is no thing."[13] This "no thing" is, of course, the very mystery at the heart of Meister Eckhart's overdetermined language of God, as the ultimate fecund nothingness from whom all determinacy comes.[14] Here Hegel's nothing—qua the dialectic of self-determining negativity—is entirely inadequate, and at the same time Desmond is right to point out that "Heidegger . . . has a truer sense of this other nothing" than did Hegel.[15]

But just *how* adequate was Heidegger's account of the nothing? In the space remaining, I will make just a few comments on what is obviously a massively complex question, not only because of the voluminous extent of Heidegger's *Gesamtausgabe*, but also because of the undeniable movement in Heidegger's own thinking over time.

Even in early Heideggerian thought, negativity has an enormously productive sense that clearly exceeds mere negation. In *Being and Time*, it emerges against the backdrop of Dasein's debt/guilt (*Schuld*). Accordingly, Dasein is defined "by a 'not' [*ein Nicht*]" and is "permeated with nullity through and through."[16] This is connected to the earlier account of anxiety. Unlike fear (which has a determinate focus), "[t]hat in the face of which one has anxiety" is the nothing: it is nothing other than "Being-in-the-world as such."[17] In his 1928 lecture course on Leibniz, Heidegger extends this sense of the nothing to the "world." The world is "not nothing in the sense of "*nihil negativum*" (i.e., just pure absence); it is rather a nothingness "that is yet something that is there [*es gibt*]," and that makes possible the appearing of

13. Desmond, *The Voiding of Being*, 106 [24].

14. See, e.g., "For God is nothing: not in the sense of having no being. He is neither this nor that that one can speak of: He is being above all being. (Sermon 62. *The Complete Mystical Works of Meister Eckhart*, 316–17.)

15. Desmond, *The Voiding of Being*, 106 [24].

16. Heidegger, *Being and Time*, 331.

17. Heidegger, *Being and Time*, 230.

all things. It is thus the "*nihil originarium*": that which allows beings to be manifest.[18]

This is a theme that develops considerably in 1929, featuring strongly in his famous inaugural lecture, *What Is Metaphysics?* Nonetheless, genuine wonder in the face of the gratuity of being struggles to emerge here, with Heidegger's focus being dominated more by a deep anxiety that is the fruit (one might suggest) of Dasein's frustrated sovereignty. "In the clear night of *the nothing of anxiety,*" Heidegger notes, "the original openness of beings as such arises: that they are beings—and not nothing." True, Heidegger is clear that "[o]nly on the ground of wonder—the revelation of the nothing— does the 'why?' loom before us." But this is a "wonder" that is the product *not* of awed *appreciation* for the given, but of oppression: "[o]nly when the strangeness of beings oppresses us does it arouse and evoke wonder."[19]

For Desmond, "erotic perplexity" is no substitute for "agapeic aston- ishment." The *eros* of the former relates to the desire to grasp the problem in determinate terms, and conquer it through the power of *logos*. It is mo- tivated by a sense of lack within, which desire seeks to fill. (This is an issue not unconnected to Heidegger's early fixation on transcendence as that which belongs to Dasein, or as Desmond might put it, "T2" as counterfeited "T3.")[20] Many years ago, Desmond noted and criticized this aspect of Hei- deggerian thought:

> At a remove from the "It is good," Heidegger thinks that *Angst* before nothingness, faced in our always impending death, reawakens us to the enigma of being. This privileging of the encounter with the absolute lack of being through the nothing marks Heidegger's thought as more erotic than agapeic. Hence

18. Heidegger, *The Metaphysical Foundations of Logic*, 195, 210.

19. Heidegger, *Pathmarks*, 90, 95. Desmond remarks that "[p]erplexity can be haunted by horror: ontological horror before the being-there of being in its excess to our rational measure" (114). Indeed. But such a sense is not to be found in Heidegger's work, at any point. The "oppression" referred to in the inaugural lecture never de- scends into a horror of frustrated perplexity in anything like the way it does in Sartre's famous reduction of the primal contingency of existence (the gratuity of being) to the mere absurd *lack* of meaning. [In *Nausea*, Sartre describes Roquentin's vision as fol- lows: "existence had suddenly unveiled itself ... leaving soft, monstrous masses, all in disorder—naked, in a frightful, obscene nakedness ... [W]e hadn't the slightest reason to be there ... these superfluous lives... Every existing thing is born without reason, prolongs itself out of weakness and dies by chance. (Sartre, *Nausea*, 127.)] Heidegger, in contrast, is far from tone-deaf to the wonder of being, and nor does he refuse onto- logical pathos (the "*passio essendi*"), even if both senses develop progressively across his career.

20. See, e.g., Desmond, *God and the Between*, 22–23.

also the privileging of futurity and the resolute will. The residues
of idealism in his version of erotic mind as *Sorge* does not suf-
ficiently guard against the "for-self" of *Dasein* seeking to assert
itself over the relativity that is always already at work. . . . This
makes it hard to grant agapeic mind . . . to see the very dynamism of lacking being as an incomplete expression of a fullness
already at work.[21]

Nonetheless, as Desmond also notes, there is a prominent thread in
later Heideggerian thought in which the openness of being is not reduced
simply to the correlation of being with the opening of being *to/in/for*
Dasein/human being (*der Mensch*); and in which wonder is more consis-
tently othered. Indeed, immediately after his critique of early Heideggerian
thought above, Desmond notes that in the later works "there is a sense of the
origin as giving, and a sense of awaiting this giving, a waiting beyond wilful
interfering . . . [that] is more reminiscent of agapeic mind."[22] How so? I end
with a few observations on this score.

In his winter 1937–38 lectures, Heidegger devotes considerable time
to teasing out what is at stake in the Greek notion of *thaumazein* (wonder),
which as Plato notes in *Theaetetus*,[23] is the *pathos* of the philosopher. Here
he argues for the necessity of moving away from more mundane senses of
wonder (to mean amazement, marvelling, admiration, astonishment, and
even awe)[24] to its more primordial or essential sense. Accordingly, to be
in wonder means that "what is most usual of all and in all, i.e., everything,
becomes the most unusual." Being is thus recognized in its irreducible
strangeness. In wonder, we come to experience and to "know," that we are
"incapable of penetrating the unusualness by way of explanation, since that
would precisely be to destroy it." But further, Heidegger adds, "[n]ot knowing the way out or the way in, wonder dwells in a *between*, between the most
usual, beings, and their unusualness, their 'is.'"[25] In this way, wonder stands
opposed to the technological reduction of being (*technē* standing "against
phusis"!),[26] which—as was suggested above—is a primary practical marker
of scientism.

Needless to say, Heidegger's engagement with Eckhartian thought
marked a decided movement away from the thinly disguised voluntaristic

21. Desmond, *Perplexity and Ultimacy*, 132.
22. Desmond, *Perplexity and Ultimacy*, 132.
23. Plato, *Theaetetus*, 155d.
24. Heidegger, *Basic Questions of Philosophy*, §37.
25. Heidegger, *Basic Questions*, 145. Emphasis added.
26. Heidegger, *Basic Questions*, 154.

agenda of the 1920s and the first half of the 1930s (with its sinister "archē-fascistic"[27] applications at one point). Thus, by the mid-1940s Heidegger is writing about "the transition out of willing into releasement [*Gelassenheit*]," with what he now calls "thinking" as "releasement to" the "open-region" of being.[28]

Nonetheless, there remains a suspicion that Heidegger's thinking of the nothing remains too "thin" to provide a robust account of the emergence of the rich profusion of being. In one of his most sustained engagements with Heideggerian thought, Desmond makes the case for that view as follows:

> Heidegger's philosophy claims to be a philosophy of finitude. . . .
> [But if] the nothing is all that there is to give the background to
> the finite foreground, then there is nothing about it that could
> *originate or create in the properly ontological respect that is re-*
> *quired by the situation. Ex nihilo nihil fit* [nothing comes out of
> nothing]. . . . There must be another sense of nothing, or another
> sense of the primal power of being as origin that cannot fit into
> this way of thinking. But this must mean that the so-called phi-
> losophy of finitude itself has to be redone.[29]

Indeed it does; and in my view, there are various places where Heidegger does precisely that. Some of the most pertinent, I suggest, are Heidegger's under-appreciated engagements with pre-Socratic philosophy (especially in the early 1940s), that address this issue of finitude, while also providing perhaps the strongest Heideggerian anticipation of what Desmond refers to as the "porosity of being."[30]

In his reading of Heraclitus in the 1940s, Heidegger introduces a strikingly open notion of being qua *physis* and *kosmos*, both of which are read as alternative names for being. As Richard Capobianco has shown, in Heidegger's reading, the motifs of "fire" and "lightning" are read in this context. Fire is "that which allows all beings to flame up in the first place," and lightning is that which "'steers' all beings into their proper place." *Physis* itself is "the pure 'emerging' that 'opens up' all beings in the first place"; it is "the inapparent joining, the noble opening up, the from-out-of-itself essencing lighting-clearing . . . [that] appears and shines forth beings as

27. Lacoue-Labarthe's apt term. See his "The Spirit of National Socialism and Its Destiny," in Lacoue-Labarthe and Nancy, eds., *Retreating the Political*, 143.

28. Heidegger, *Country Path Conversations*, 70, 80.

29. Desmond, *Art, Origins, Otherness*, 250, 252.

30. For a detailed account of some aspects of this theme, see my, "Heidegger on (In)finitude and the Greco-Latin Grammar of Being."

a whole." The "shimmering" *kosmos* is another name for this "primordial emblazoning-adorning."[31]

However, more remarkable still is Heidegger's account of Anaximander's *apeiron*,[32] again presented by Heidegger quite explicitly as another name for being.[33] Focusing especially on two famous Anaximanderian fragments—including: "the beginning [*archē*] of beings is the limitless [*apeiron*]"—Heidegger links this to the quite ubiquitous motif in his own later work concerning the granting/withdrawal movement by which the presencing of beings is made possible. Accordingly: being is the unlimited superabundant *archē* from which beings step forward, from which they are granted determination or "contours" in which they persist for a while, and by which they become concrete and limited beings. But all beings inevitably (Anaximander's "necessarily") recede again to their source, giving up the limitation of their contours, thus returning to the unlimited. This *archē*, this primordial *Abgrund*, this nothing (perhaps also this "*nihil originarium*") is beyond all ontic causation. It richly grants, even as it remains concealed behind the givenness that it releases.

Of particular note here is the giving itself. On Heidegger's account, the *archē* is not surpassed in the granting of determinacy (it is the utterly unsurpassable); it is rather an "enjoining egress [*verfügende Ausgang*]." Beings remain always enjoined to their source. As such, the *archē*, being, is the "between [*Zwischen*]."[34] This theme of enjoinment reiterates the same motif in Heidegger's reading of Heraclitus: "*phusis* as the essential jointure (*harmonia*) of emerging and submerging (self-concealing) in the reciprocal bestowal of its essence."[35] In this constant *enjoining* of beings with their unlimited super-abundant ground, there is a powerful sense of what Desmond calls the *porosity* of being. The *a-poria* of mere perplexity is thus overcome insofar as agapeic mind comes to dwell, with wonder, in the "enjoining egress" of being. Further, the privative construction of *a-peiron* should not be understood as a mere negation of limit (any more, Heidegger suggests, is the case with *a-lētheia*, which is no mere matter of the negation of covering).

31. Capobianco, *Heidegger's Being*, 12–13. See Heidegger, *Heraclitus*, §8.

32. There are various places where Heidegger discusses the Anaximander fragments. Perhaps the most vivid are his 1932 and 1943 lecture courses, respectively: *The Beginning of Western Philosophy* (*Anaximander and Parmenides*); and *Basic Concepts*.

33. Heidegger is explicit on this point: "We must now no longer be content with the introductory characterization according to which Being is appearance. That is not wrong but is insufficient; the essence of Being is to be understood on the basis of the *apeiron*" (*Beginning of Western Philosophy*, 25).

34. Heidegger, *Basic Concepts*, 93–94.

35. Heidegger, *Heraclitus*, 107 (and §7).

As he puts it elsewhere, this nothing "as the nothing of beings, is the keenest opponent of mere negating."³⁶ For the *apeiron*—being, qua the unlimited nothing from which all beings appear in their determinative contours—is a "not" that *makes possible* an unbounded richness.

Desmond's reserve around Heidegger's apparent "postulatory finitism" is well taken. Yet there are resources within his later work that would seem to be pointing in quite different directions, even if great care needs to be exercised when dealing with his textual subtleties. Still, there is an increasingly rhapsodic tone to Heideggerian thought through the 1940s and beyond, in a way that bespeaks something like agapeic mindedness. He writes, for instance, of "the poverty of reflection" that nonetheless allows dim access to "the promise of a wealth whose treasures glow in the resplendence of . . . the inexhaustibleness of that which is worthy of questioning."³⁷ Such thinking has an essential kinship with thanking: "*Denken/ Danken.*"³⁸

Reclaiming Ontological Porosity, beyond Scientism

The natural sciences are right to deal with the domain of beings in their determined properties and causal relations, for this mode of methodological openness to being provides legitimate and important insight. The error of scientism—and it is a deep one—is the delimitation of all thinking to one such particular "projection of being" as the early Heidegger put it. Scientism amounts to a decided narrowing of thought, and thus to its impoverishment. It is to foreclose on the very possibility of thinking the larger overdetermined context within which all inquiry occurs, and to impose the circumscription of all thinking within a closed circle of immanence. It is to substitute scientific curiosity for *all* other modes of thought, and thus to undermine them, be they metaphysical, ethical, spiritual, or aesthetic. But yet Desmond is correct: science itself is only possible because of the porosity of being that grants the abundant whole to be investigated, and indeed the potential for understanding it. In declaring itself supreme, scientistic thinking forgets that it is parasitic on the very power of thinking that is "derivative from this original porosity."³⁹

---

36. Heidegger, *Off the Beaten Track*, 85.

37. Heidegger, *The Question Concerning Technology and Other Essays*, 181.

38. Heidegger, *Pathmarks*, 236.

39. Desmond, *The Voiding of Being*, 99 [18].

Bibliography

Bacon, Francis. *The Works of Francis Bacon.* Edited by James Spedding. Vol. 6. New York: Garrett, 1968.

Capobianco, Richard. *Heidegger's Being: The Shimmering Unfolding.* Toronto: University of Toronto Press, 2022.

Colledge, Richard. "Heidegger on (In)finitude and the Greco-Latin Grammar of Being." *Review of Metaphysics* 74.2 (2020) 289–319.

Desmond, William. *Art, Origins, Otherness.* Albany, NY: SUNY Press, 2003.

———. *Being and the Between.* Albany, NY: SUNY Press, 1995.

———. *God and the Between.* Malden, MA: Blackwell, 2008.

———. *The Intimate Strangeness of Being.* Washington, DC: Catholic University of America Press, 2012.

———. *Perplexity and Ultimacy.* Albany, NY: SUNY Press, 1995.

———. *Philosophy and Its Others: Ways of Being and Mind.* Albany, NY: SUNY Press, 1990.

———. *The Voiding of Being.* Washington, DC: Catholic University of America Press, 2020.

Eckhart. *The Complete Mystical Works of Meister Eckhart.* Translated by Maurice O'C. Walshe. New York: Herder & Herder, 2010.

Hawking, Stephen. *A Brief History of Time: From the Big Bang to Black Holes.* Toronto: Bantam, 1988.

Heidegger, Martin. *Basic Concepts.* Translated by Gary Aylesworth. Bloomington, IN: Indiana University Press, 1993.

———. *Basic Questions of Philosophy.* Translated by Richard Rojcewicz and André Schuwer. Bloomington, IN: Indiana University Press, 1994.

———. *The Beginning of Western Philosophy (Anaximander and Parmenides).* Translated by Richard Rojcewicz. Bloomington, IN: Indiana University Press, 2015.

———. *Being and Time.* Translated by John Macquarrie and Edward Robinson. Oxford: Blackwell, 1962.

———. *Country Path Conversations.* Translated by Bret Davis. Bloomington, IN: Indiana University Press, 2010.

———. *Heraclitus: The Inception of Occidental Thinking.* Translated by Julia Assaiante and S. Montgomery Ewegen. London: Bloomsbury, 2018.

———. *The Metaphysical Foundations of Logic.* Translated by Michael Heim. Bloomington, IN: Indiana University Press, 1984.

———. *Off the Beaten Track.* Translated by Julian Young and Kenneth Haynes. New York: Cambridge University Press, 2002.

———. *Pathmarks.* Cambridge: Cambridge University Press, 1998.

———. *The Question Concerning Technology and Other Essays.* Translated by William Lovitt. New York: Harper and Row, 1982.

Kant, Immanuel. *Critique of Pure Reason.* Translated by Paul Guyer and Allen Wood. Cambridge: Cambridge University Press, 1998.

Lacoue-Labarthe, Phillippe, and Jean-Luc Nancy, eds. *Retreating the Political.* London: Routledge, 1997.

Plato. *Complete Works.* Edited by John M. Cooper. Indianapolis: Hackett, 1997.

Sartre, Jean-Paul. *Nausea.* New York: Penguin, 1965.

8

The Determinations of Medicine and the Too-Muchness of Being

Jeffrey Bishop

Introduction

MEDICINE IS NOT A science, *per se*. Certainly, medicine draws on knowledge gained through scientific inquiry. However, it is not a science, if by science we mean that process or methodology that seeks to describe or explain the hidden causes or forces that animate the living human body. It is sometimes referred to as an applied science, but by that we really only mean that it is applied knowledge. Medicine is a practice; it is not about discovery or even about the body of knowledge. It is about taking the generalized knowledge—the body of knowledge—gained through the scientific method and attempting to apply it to particular instances. Surely, there are physicians that are also scientists. But they must know when they are seeking knowledge of the world of humans, and when they are seeking to manipulate the world of human bodies and psyches. Thus, medicine does not articulate theories that cover multiple particulars, the way that science does, even while it utilizes theories and models derived scientifically. Still the particular patient encountered is never fully covered by the general laws, or the summation of all laws, or the grand model comprised of all the relevant models of human bodily function. Scientific knowledge is fractured, infinitely perspectival (the bad kind of infinite), and will always be changing, since there is no final synthetic whole that could cover the being of all human beings. After all, the being of a living patient exceeds all general knowledge. A single human being, in her being, is in excess of even our grand synthetic models. Science is of the general, it seeks a covering theory for multiple instances of what is presumed to be similar phenomena. Medicine is of the particular.

That means that medicine's activity starts from a place of determinate knowledge synthesized together from all the pieces of information about human beings derived from the scientific method. Its knowledge seems to be settled in that it imagines that all physiological processes obtain in every instance in the same way: all hearts operate according to the Frank Starling theory of cardiac contractility; all mean atrial pressures are the product of cardiac output and systemic vascular resistance; glomerular filtration is an activity accomplished in the kidney to filter toxins from blood into the urine, and the rate of glomerular filtration can give a snapshot of the effectiveness of kidneys at accomplishing this task.

In the absence of mathematical models to describe normal or pathological biological functioning, medicine resorts to propositional description. Examples might be: the classical symptoms of acute pancreatitis include epigastric burning sensation that radiates through to the back, along with nausea and vomiting, all of which is made worse with eating, and along with elevated amylase and lipase levels in serum; the criteria for antisocial personality disorder include failure to conform to social norms, deceitfulness, impulsivity, aggressiveness, reckless disregard for safety of others, consistent irresponsibility, lack of remorse; Becks Depression Inventory is a list of symptoms that can be scored to give a sense of the different degrees of depression, from mild to moderate to severe. Medical knowledge is determinate knowledge, knowledge that assures the physician that she has mastery of all the possibilities as the indeterminate being presents itself to her.

However, careful attention to the patient inevitably shows the failure of this determinate knowledge; in each instance, in fact, the patient will exceed the sum of all medical knowledge, even if it is effective. In fact, the most memorable patients (or cases as they are sometimes called) are those that defy determination, those patients whose illnesses do not conform to the determinations. Where William Desmond, in his essay on astonishment in science, describes the movement of the human encounter with reality in science from the overdeterminate excess of being, to the indeterminate provisional knowledge of particulars, to determinate knowledge of theoretical kinds, medicine proceeds in the opposite direction. It moves from determinate to indeterminate to overdeterminate being, from curiosity to perplexity to astonishment. But too often, medicine arrives too late at the mystery of being, usually only when something has fallen out of being. The question for medicine then is how to let the overdetermination of being inform its determinate practices, and how to engage human beings without overly circumscribing living being, or constraining it to the point of stifling it.

I shall begin by giving two patient descriptions that illustrate the way that determinate medical knowledge falls short of human being; in other words, wise medical practitioners understand that determinate being as it is imagined slips into indeterminate being. I will also explore the kind of openness that is necessary so as to keep the determinate knowledge from constraining its practices, and from foreclosing on and circumscribing the overdeterminate reality of each actual patient.

The Curiosity That Is Medical Thinking

Case 1: A Depressive Patient

In the Autumn of 2003, a fifty-year-old woman (whom I shall call Emma) came to my clinic with multiple bodily complaints. I had been her primary care physician for a few years before this visit. She complained of back pain, headaches, and fatigue. She also noticed a loss of appetite. However, her weight was unchanged. She said that occasionally she would have some bloating and abdominal pain. When further questioned, the abdominal pain was not so much pain as distension, with perhaps some associated nausea. She also noticed her heart racing after minimal exertion, and at night she noticed her heart pounding in a way that she had never noticed before. Emma denied shortness of breath or chest pain. She had hypothyroidism, which had been well-managed with oral thyroid hormone replacement. And she had noticed less frequent menstrual periods, and somewhat heavier flow of blood when she did have periods. I did a physical exam on her and found nothing out of the ordinary.

I ran several blood tests, including a complete blood count (CBC) and electrolyte and liver panels, as well as a test for her thyroid hormone levels, and scheduled a follow up appointment for two weeks. I ordered an electro-cardiogram, which was normal, as well as an outpatient heart monitor, which she could wear and trigger if she had the sensation of racing heart. I also asked her to keep a diary of her headaches and back aches for us to review at her next visit. Her ECG and blood work all came back as normal, and within a few days, the cardiac monitor would come back showing normal sinus rhythm with occasional episodes of sinus tachycardia during the times that she reported a racing heart. Like her physical exam, her laboratory exams were normal, as were all her cardiac tests, suggesting that nothing major was going on.

I also happened to know that Emma had worked as a mid-level HR manager in a large company, and that she was in charge of hiring and firing employees, as well as intervening in interpersonal conflicts within the company. It was a high-stress job, where she had a lot of responsibility and very little authority to make decisions on her own. She was also dealing with a sixteen-year-old daughter who had been dabbling in drugs and alcohol, and had been known to skip school on occasion. All of these stressors and her worry over her daughter's behavior had made home life stressful and resulted in several sleepless nights.

Upon Emma's return to see me, we reviewed her symptoms, which had not worsened or improved in the two weeks. Upon reviewing her diary, it became apparent that the headaches were not that frequent and usually came after a restless night, and were associated with fights and tension in the family. Emma's story is not that uncommon, though the particulars might be different from person to person. She was likely suffering from depression, but curiously, she had no symptoms that one typically associates with depression. She denied sadness, or tearfulness, or anxiety. Emma was shocked to see that her headaches were associated with sleepless nights, which were associated with the stress arguments with her daughter. People are often unreflective about their own feelings, sometimes using their rational minds to evaluate the situation, but not permitting it to examine or even notice the feelings associated with the situational features.

I said to Emma, "Whenever someone has multiple complaints that don't fit an anatomical or physiological pattern, I sometimes wonder if the person might have depression. Do you think you might be depressed?" Emma immediately began to cry and to protest that she was not depressed. The incongruence between her protestation against my suggestion that she might be depressed and her tearfulness led me to conclude with even more certainty that she was in fact depressed. I paused for a moment and gave her time to compose herself, and then proceeded to ask questions from the Beck's Depression Inventory, a series of questions that would assist in diagnosing depression, and might give us a sense of the degree of depression. After finishing the questions, Emma scored in the mild range of depression.

When I told her the results, she became tearful again and shook her head, again in protestation. We sat there quietly for a moment; it is sometimes best to let the patient say something to break the quiet. Suddenly she blurted out, "Dr. Bishop, do not look at me that way; you look like you know something about me that I do not know about myself, and I do not like it!" I was taken aback, and I was indeed embarrassed. I did not think I had given her a look, but apparently I had. For the first time in my career, I had been caught deploying what Foucault had called "the gaze." That is not to say that I had not deployed the gaze in the past, for all doctors deploy the gaze of determinate knowledge. It was just the first time I had been caught looking at someone with the gaze, who was protesting against it. I am sure that I was confident in my diagnosis, more so after her tears, and that I had given Emma that knowing look. It was indeed a look, a kind of looking at a patient as if looking at them through a specific lens. It was a knowing look, determined by my ready-made concepts of depression given to me by the Diagnostics and Statistics Manual of the American Psychiatric Association. It was a look born of the determinate knowledge of medical expertise.

Case 2: A Septic Patient

In July 1994, I was the resident—postgraduate—physician in charge of the Medical Intensive Care (Therapy) Unit (MICU). I had been called around 7 PM to see a twenty-one-year-old woman (whom I will call Charlene) in the Emergency Department. She had arrived in the ED in the very early morning of the same day complaining of a hot, swollen upper thigh. She had been evaluated during the early morning, and admitted to a colleague's medical team around 7 AM. She was diagnosed with cellulitis, an infection typically caused by a bacteria called streptococcus. Streptococcus is a ubiquitous organism. In fact, we all have some streptococcus on our skin at this very moment. Sometimes, this bacteria can slip beneath the exterior skin barrier into the subcutaneous tissue, resulting in a diffusing infection that runs through this subcutaneous tissue. In Charlene's case, the infection began in her upper thigh, where she had some pain and redness. My colleague had evaluated her and written admission orders for her to be placed on the medical floor of a busy, publicly funded hospital in a large metropolitan city. She was to receive IV antibiotics. However, the hospital was full and the admission orders, which had been entered at about 9 AM, were still on her chart at 6 PM that same day. She had not been moved to the medical floor, nor had she received the IV antibiotics.

Charlene was a college junior, working a summer job as a part-time administrative assistant. She had done well in high school, but was struggling a bit in college. I do not recall her major area of study, but I do remember that she was having second thoughts on what to study. For some reason I do remember that she was into fashion, which is an odd thing to remember, nearly thirty years after her death. Other details were hard to come by in my interview with her. She was a bit groggy and was having some respiratory difficulty. She was in a sepsis syndrome, and we were going to have to make some rather difficult decisions over a very short period of time.

Her blood work was really out of sorts. Her creatinine, an indicator of kidney function, had been normal in the early morning hours, but still had signs of intravascular volume depletion. This condition is sometimes referred to as "dehydration," but physiologically speaking she did not have a total body water deficit. It means that there is less blood volume, which means her heart has to work harder and faster to maintain the same blood pressure. Her blood vessels were leaky, in that protein, electrolytes, and white bloods cells, as well as plasma, were exuding into her interstitial tissues. Her white blood cell count was elevated to about 15,000 with a neutrophil shift, suggesting acute bacterial infection. Given the redness of her leg and all the laboratory findings, everything pointed to cellulitis.

By the time I saw her around 7 PM that evening, her creatinine had increased, meaning her kidneys were beginning to fail, and her WBC count was even more elevated. In addition, her blood pressure was low and her heart rate was very high. She was also acidemic, which means that she was not perfusing her body very effectively due to the lower blood pressure; some parts of her body were switching from aerobic metabolism (oxygen-based metabolism) to anaerobic metabolism (metabolism that occurs when there is less oxygen availability). Her respiratory rate was increased in an attempt to get more oxygen to her body and in an attempt to reduce the acidemia. But the fact that her blood vessels were leaky meant that more fluid was exuding into the interstitial tissues in her leg and there were signs that this process was also beginning in her lung, making it harder for oxygen to move from the air space in her lungs to the blood space. That in turn means that less oxygen was getting to her extremities, her kidneys, and her brain. She had worsened in those few hours. I asked the nurse to give an IV dose of penicillin immediately, and I began IV fluids to try to increase the volume in her blood vessels to get her blood pressure up.

When I examined her leg, it was red and tense. My colleague had said it was much worse than earlier in the day when he had seen her. Her groin and upper thigh was very tense. I called a surgery resident, who was more senior than I, asking her to see Charlene because I was worried that the infection was not just in the superficial skin. I was worried that the infection had gotten into her muscle. Muscles are covered in a thick fibrous sack called the fascia. When the infection gets in the muscle, the inflammatory process and infection puts more fluid inside the muscle, so much so that the pressure inside the fascia increases to the point where even blood cannot get to the muscle itself. Instead of cellulitis, I was worried she had necrotizing fasciitis. If I was correct, she would need immediate surgery to clean out the infection and relieve the pressure inside the muscular fascia.

We moved her to the MICU, and over the next few hours we would need to add pressors (medications that will increase blood pressure) to try to increase blood supply to her failing kidneys and to other vital organs. Over a very short period of time, it became clear that we would need to place a tube into her trachea and to place her on a ventilator in order to take over some of the work of breathing for her; she was not going to keep up. Charlene's bodily functions were spiraling out of control.

As my intern prepared for the intubation, I went over to Charlene and began to explain what was going on. I got close to her and explained that her whole body was in crisis and that we were going to have to help her body breathe by placing the tube in her trachea so that we could push oxygen into her lungs, and so that she could blow off the carbon dioxide. As I got closer to her, she became more awake as I looked into her eyes. As I gave her the information, it was as if she could see into my mind, could see the concern I had, the fear I had that she might die. In fact, in retrospect, I think she saw that in my mind her death was virtually assured. It was not anything that I said. I said all the right things, exuding confidence, framing the gravity her situation with facts about what we were doing, all of which were said in hope. Astonished, she looked back at me, somehow knowing what I knew; I did my best to let her know that I would stand with her, but who will not fall short before the abyss!?! She looked back, in hope, that somehow all that I knew would be enough to stop her from what my eyes were communicating. Seeing her now, in my mind's eye, I marvel at what was lost that day.

By the time the surgeon got to the bedside, Charlene was on the ventilator, getting high dose antibiotics and pressors. The surgeon examined her and decided not to take her to the operating room (OR). I should have, at that point, insisted that the surgeon flay open her leg, preferably in the operating room and if not there, then right there in the MICU. To this day, I still do not know why the surgeon didn't debride her leg right there. Short of that, I should have flayed open her leg myself right there in the MICU, but I was too inexperienced to be so bold as to do that over the wishes of a surgeon of higher rank. Wisdom is the thing you get just after you need it. I have never failed to be bold in my conversations with others—doctors or scholars—since that day.

Charlene continued to worsen. By 10 PM, she began to show signs of worsening renal failure. By midnight, she went into cardiac arrest. During the cardiac arrest, we sent blood work. The results came back too late. Because of the necrosis in her muscle, a massive amount of intracellular potassium had been released into her blood, which caused her cardiac arrest. I was unable to get her heart to start back. At 12.30 AM, less than twenty-four hours since her arrival, I pronounced Charlene—a twenty-one-year-old woman—dead. About two hours later, I was paged by someone in the laboratory stating that her blood cultures were growing gram positive cocci in chains. The next day they were identified as Group A beta-hemolytic strep. Six weeks later, her autopsy showed that she had necrotizing fasciitis.

Determinate Thinking in Medicine

Medical thinking begins in ready-made concepts born of the determinate thinking of medical science. It is a practice that launches from a social location, where the concepts are policed through the institutions of medical education, professional societies, and standards of care, and these concepts enforced by economic and legal apparatuses that ensure that the practices are repeated through time. These standards of care, these ready-made concepts become the measure against which practitioners are judged, and by which they come to police themselves.

These ready-made concepts and models of the living body work for the most part and most of the time. Yet, these concepts and models fall short of the reality of living beings in that they are inadequate to the flux of living flesh and the flow of time in the unfolding of life itself. Medicine deploys these ready-made concepts for good, and it is for this reason that people seek out medical practitioners. The ready-made concepts and models are represented symbolically through equations. For example, mean arterial pressure is equal to the product of cardiac output and systemic vascular resistance [MAP = CO * SVR]) or through graphic representations such as the Frank Starling Curve (see graph below), or through a set of propositions, such as the propositional definition of antisocial personality disorder put forward in the Diagnostics and Statistic Manual of the American Psychiatric Association. The multitude of factors that make it possible for a living being to continue living can be represented for the most part and in the main with these symbolic representations, or through models of biological and physiological activity. These models are representations of a kind of determinate thinking, representations of living being, which is itself in excess of the various ready-made concepts and models. At some level, medicine must needs have this kind of determinate thinking. It is from the scaffolding of concepts that medical practitioners launch into the flux of human biological and physiological being.

Painting by Number: Medical Representation, Perplexity, and the Indeterminacy of Being

If one imagines all the various ways that one can piece together a model of the functioning body or a functioning psyche, one is imagining something very different from a living body and a living psyche. As we have seen, one would have to piece together a myriad of perspectival snapshots. Is it possible to have a view of something from every perspective? Every single scientific perspective from the subatomic, to the atomic, to the molecular, to the cellular, to the physiological, to the organismal, to the social level gives us some view? Could we synthesize a view from every single experiment at each of those levels? Would we synthesize a view from every possible perspective? The answer is, of course, no! An exhaustive model of the human body and the human psyche is impossible. After all, there would be an infinite number of perspectival points from which one could examine any object, not to mention an object like the functioning human body, which is in motion, let alone a human psyche—itself in motion—that is altogether different from the object of the human body. Thus, the determinations, infinitely perspectival—a bad infinite—though they may be, would still not be exhaustive of the being of the whole and would fall short of the flux of the human body and the human psyche.

These various snapshots of reality, which are the determinations of medical science, have to be reconstructed into an image of the whole. When thought of in this way, it seems obvious, as we saw with Emma, that my diagnosis would fall short of Emma's lived experience. Beck's Depression Inventory gives a determinate description—an image or even a drawing, if you will—of what depression looks like. It is a symbolic representation of an idea—an imagined picture of "depression" drawn by scientific medicine. Its status as a representation is roughly on par with a painting from Pablo Picasso's Blue Period. While the abstraction that is Beck's Depression Inventory converts qualities into scoreable quantities, it is even less accurate to the experience of depression than Picasso's The Blue Room or The Old Guitarist.

The medical models, because grounded in the determinate thinking of medical science, are sometimes assumed to be better representations of the flux of subjective psychological being. The illusion created by quantitative representations permits a kind of self-assured smugness as practitioners cast their gaze upon the patient. The gaze, because originating in the powerful determinate thinking of medicine, asserts itself with a certainty in excess of the patient's folk imagination about her condition. In fact, because of the powerful location of medical practice, patients will even question their own take on their own lives and readily embrace the ready-made concepts of medical science.

When I saw Emma, I immediately began to size up her problem. I had placed the abstract, statistical artist's representation known as Beck's Depression Inventory alongside Emma's life. The image she had of herself—her self-understanding—did not match-up to the medical image of a depressed person. She felt the oppressive weight of the determinate knowledge of psychiatric medicine, and she resisted my deployment of that gaze onto her body and psyche.

In time, I have come to trust Emma's (and other patients') own self-understanding, and to understand the interpretative shortcomings of medicine, rather than to trust the numbers of the statistical artists, who paint by number, images that might be correct, but have very little resemblance to the truth of human subjective being.

Physicians and surgeons may say, "Well, subjective interpretation of one's own psychological state may be perfectly legitimate, but those of us that deal with the physical entities of human being—like hearts, lungs, muscles, infections—our determinations are more certain and secure. It just would not do for some 'subjective' interpretation to govern medicine's activity." All cardiovascular systems—indeed life itself—operates on the same inviolable laws of physics: MAP = CO * SVR; the stretch placed on cardiac muscle increases the efficiency of contraction; glomerular filtration is a function of cardiac output. These statements are true in the general, but how they play themselves out in particular instances depends on the life history of a patient.

For example, a patient that had rheumatic fever with cardiac involvement, might have undetectable micro-injury of the heart muscle such that her heart behaves in a way that teeters on the edge of vascular resistance in a way very different from a normal heart. The particular heart may behave in a way that exceeds what the typical, determinate (scientific) description can capture in its general terms.

Charlene's death was a tragedy with several errors along the way. Were these errors in scientific knowledge, a failure of science not having enough information? It seems not. Certainly, she had cellulitis; my colleague got that right. This determinate knowledge was sufficient. Certainly, at some point what had been a cellulitis transformed into necrotizing fasciitis; I got that correct by deploying the determinate of medical science. My surgery colleague was certainly incorrect; so it was not a failure of scientific knowledge, but a failure on the part of a particular physician to recognize what was going on. Thus, there are two kinds of errors seen in medical practice. One would be the error of insufficient knowledge in the scientific community;

the other would be a failure of an individual physician in recognizing something that could have been known. Yet, there is another kind of error in medicine to which Samuel Gorovitz and Alasdair MacIntrye brought our attention, and of which even Aristotle was familiar. All knowledge is of the general; the particulars of a reality, which are in motion, exceed the snapshot of generalized knowledge.

In Charlene's case, the temporal dimension of the unfolding of a disease process made it difficult for us to see—and to see fully only in retrospect—where we might have gone wrong. My colleague may have been correct that all the symptoms described by Charlene and all the signs manifested in her body, fit very tightly to the diagnosis of cellulitis. My colleague may not have in fact been wrong. However, the snapshot description of cellulitis and the snapshot description of necrotizing fasciitis are insufficient in that the cellulitis can spill over into a fasciitis without a clear line of demarcation between the two, and the inflammatory response can spill over from fasciitis to system inflammatory to sepsis syndrome without clear demarcation. As with living being, disease processes are flowing, fluid, moving, resistant to the ready-made concepts of determinate medical science.

By the time I saw her at 7 PM and—given that by that time she had necrotizing fasciitis—there was probably nothing that could be done. Necrotizing fasciitis is virtually always deadly when it occurs in the larger muscle groups. Would she have lived if at 7 AM she had gotten immediate surgery? Probably. Would she have lived if at 7 AM she had gotten the IV Penicillin that she only got around 7 PM? Possibly. The flux of the bacterial beings that had seeded Charlene's upper thigh muscles, one or all among three muscles (the psoas major, sartorius, or adductor longus muscles), combined with the flux of her body's inflammatory response, combined with the flux of her own cardiovascular system means that no determinate set of descriptions and synthetic reconstructions will ever be sufficient to the reality of being, or in this case, the falling out of being, even amongst the more stable entities like bacteria and bodies.

Thus, whether we are dealing with the entity of the body or the entity of the psyche, the being of humans is indeterminate in relation to medical knowledge, however stable we may hope medical knowledge is. What is astonishing is that the being of humans—body and soul—are fluid things, things in act that slip past the determinations. At best, medical knowledge is provisional, temporary, and always falling short of the objects of medical concern. The result is a kind of astonishment at the instability of human biological and psychological entities, revealing a too-much-ness of being human, and leads to the astonishment that there is life at all.

Overdeterminate Being and the Astonishment
of Falling Out of Being

The Porosity of Being and Knowing

For me, as the doctor, Emma's case was a cut and dry case of depression, likely of situational origin. Yet, this diagnosis was not anything that she recognized about herself. The mind is a powerful sort of thing; it is an odd thing, so odd that it even resists being called a "thing."

For the doctor, the mind of a patient is read as if it is a surface phenomenon. A patient suffering depression or anxiety might show observable signs of increased heart rate, inability to focus, tearfulness, rapid breathing, sweatiness, or trembling. But these observed phenomena are seen perspectivally from outside the body or from outside the "internal" experience of the patient's body; but still they are read from the "surface" of the patient's body and thus could be signs of virtually anything. In fact, many "organic" diseases have just these surface "signs." Even Charlene had a racing heart, clamminess of her skin, rapid breathing. The doctor is without access to the "interior" world of the patient's sense or observation of these bodily manifestations from within her own body.

And yet, I—as the external observer—could enter "into" her consciousness through conversation, through imagination, and even could feel her tearfulness, her sadness, the solidity and weightiness of her life, all of which she herself denied feeling. And she felt the weightiness of my gaze, the burden of which was too much and pressed in on her to the point of breaking. Whatever burden her body and mind might have been under—only part of which she seemed to be aware—the burden of my gaze, itself weightless, broke her even more. She felt my feeling of her feeling, which increased her burden; the weight of my gaze and my own feeling brought to the surface of her consciousness the weight of her situation, which had all along been there in the depths of her body, only surfacing for her in the moment. The gaze, which seemingly has no materiality, weighs heavy upon her body and her consciousness.

Yet, she seemed to ask of me, with her tears, "Why should your gaze, your interpretation, be more 'real' than my own? Why should I bear it in the way that you—with your determinate knowledge—want me to bear it?" And I felt the burden of her question, indeed her accusation, on me. I had placed upon her the burden of my gaze and she in turn placed the burden back onto me. I was—indeed I am—embarrassed that I had increased her burden through my clumsy deployment of the medical mind-set, with its ready-made concepts, which, though seemingly weightless, carry a weightiness.

The weightless weightiness of my ready-made concepts seemed to bear down on her—pressing her downward, de-pressing her with the weightless weightiness. What I thought was merely an image placed alongside her body and mind, had penetrated into her body and her mind unexpectedly, which was a kind of violation—an unintended violation, but a violation, nonetheless. In fact, I *had not* been looking through a lens at the surface manifestations at all; in fact, my gaze had entered her body into her mind, without me realizing it. I had lost control of the weightless weightiness of my gaze, and it entered through the porosity of her being and her becoming, adding weight, de-pressing her from the "inside." Who can bear these concepts that slip, unbeknownst to the observer, from the mind through a look, passing by the surface of her body into the depths of her own awareness?

Even Charlene experienced the weightiness of the weightless knowledge I possessed, when astonished, her eye caught my own in the moment before I had to put her to sleep, to put her on a machine. I had communicated to her that she was dying, not with my words, but with a look and she seemed to move out of the stupor induced by her body into a state of knowing that she may not wake up, the mind, being tethered to the body as it is.

Astonishment at Falling and at Being

Certainly, the mind is tethered to the body. Yet, it is capable of reaching well beyond the body to the body's past, or her own past psychological experience. Through imagination, it is capable of leaping into multiple futures exploring them—however limitedly—for possible problems or possible hopes or possible fears. Even within the locality of the body, the mind does not seem to localize; my awareness is ecstatic—standing out—from the body into the room in which I sit, my middle daughter writing an article on a new film-release across the table from me, my youngest daughter upstairs singing, strumming her guitar, and my oldest daughter downstairs in her room, her own mind racing from the mania that plagues her.

But then, on the other hand, the mind also *does* seem to localize in the body, here into my fingers as I type, now into shoulders' soreness from yesterday's work, and then to the healing wound on my leg. The healing wound was there all along, yet not totally in my awareness until this moment; but still it was there, lingering in the hinterlands of my awareness, until my awareness entered fully into the wound.

The body can deceive the mind. After all, the body retains memories—memories that are often not accessible to the mind. Trauma gets lodged in the body, sometimes confounding the mind. A person who has suffered sexual abuse may have lingering, nondescript pain in her pelvic floor that only rises to the level of her awareness from time to time. To the mind, the pain is not associated with the trauma, which she may even seem to have forgotten, though no experience is every truly forgotten. The mind seems so powerful, but it is possible for the mind to be forgetful of the body, or to have forgotten what the body has suffered. The body may have its reasons for pain, of which reason knows nothing. The body can deceive the mind, "remembering" the pain, without revealing its memory to the mind. The body may deceive the mind, but it may be a deception the mind participates in.

After all, the mind can deceive the mind. Perhaps it was the mind—it is difficult to say for certain—that has hidden the experience in the body, banishing the memory to a locality within the body. By banishing the memory from awareness to a locality within the body, the mind seems to imagine it can free itself from the burden, by having the memory localized to one segment of the body. Perhaps, there in the pelvis or the chest the weight of the burden can be more easily born, managed. Thus, a traumatic memory might have been in the consciousness, only to be placed into the body, banishing a memory from memory, such that it doesn't even recognize the memory at all. Or perhaps the memory only arises from the localized body to the non-localized consciousness under the duress of the circumstances.

The mind can even deceive the body. How often, as a child playing amongst the shadows of the trees, did I look down to see blood dripping from my hand or knee, not knowing where or when or how the injury happened. In fact, there was no pain at all until I looked to the injury, mind rushing to body to feel it. And I hear my cousins calling out to me, my mind pushes the pain from my body, even willing the body to forget the pain in my hand to grab hold of the branch, my cut hand screaming in pain to my mind, my mind telling my hand to grab hold of the branch firmly, telling my body it will be ok. The mind commands the body to ignore the pain, telling the body to do things in spite of the pain, despite the pain, indeed even to spite the pain.

There is then a complex intersection of body-mind and world, where in the medical encounter the "world" is another body-mind. The motion and flux of body as body and mind as mind, and body as mind, and mind as body along the two vectors is complex, and if one pays careful attention, one sees that one's careful attention is insufficient to the flux of the body-mind of the other. Moreover, each body-mind of the other is in relation to a world of matter and energy, and culture and politics, and institutions and technics, that both constitutes it and exceeds it. Thus, the intersection is not a simple cross of lines, but a node of mutual influence of beings flowing and mingling together, mutually influencing one another.

My medical knowledge (ready-made concepts) flows from my mind through my body (the gaze) and then through Emma's perceptive and interpretative apparatus, changing her; and she in turn returns the question to me, throwing my ready-made concepts into question, asking me to perceive her and her life from the "inside." The fact that she felt a burden placed on her by my gaze and the fact that I perceived her burden—the burden I had placed on her with my gaze—suggests a kind of porosity of being. And if that porosity of being could result in a kind of weighing down of her world by my ready-made concepts, then it is also true that the porosity of being lends itself to the possibility that my actions and my taking up with her being might lighten her load, if only I move along *with* her, attuning myself to her, instead of burdening her.

It is this lightening of loads that is the work of true healing. But medicine's practitioners are not habituated at entering into the flux of being of others in this way. They are practiced only at analysis, which arrives too late to the flux of bodies and minds. They are practiced at conceptual synthesis of medical models, but not at launching from the models of determinate knowledge into the flux of being. Yet, sitting right there in front of them is a body-mind in all of its complex motion in relation to the whole of being, and then there is the final realization that neither the patient nor the doctor is in final control.

It was there at the brink of an abyss—where control is impossible—that I stood with Charlene, at an intersection of all the failures of knowledge before the flux of being, and of all the failures of living being to sustain itself against the flux and chaos on which life ultimately rides. For a moment, Charlene and I stood united in astonishment before the possibility of not-being, a possibility that held a different kind of import for Charlene than for me, at least in that moment. And I ache that she is no more, astonished that something so substantial and alive, someone with whom I was held in a momentary entanglement, and who transformed my being and stays so vividly with me even now, no longer is. And I find myself reliving this memory, once again, arrested in astonishment at the hope that she will be again one day, and the hope that no thing and no one is lost in the excess that is the Absolute.

9

Attending to Infinitude

LAW AS IN-BETWEEN THE OVERDETERMINATE AND PRACTICAL JUDGMENT

Jonathan Horton

In between modalities

Law's province lies in between Desmond's modalities of wonder and the starkly determinate; between the high horizon of justice and the often unwonderful. High conceptions of justice are brought to bear at the point of judgment: the verdict, the judge's ruling or order, the "enactment" of legislation, or the making of an executive decree or order. With each of these blows, the overdeterminate and indeterminate imperatives of justice are enacted upon actual circumstances, in a marked decisive act.

The circumstances that call for judgment might be *past or present* facts (a killing, a breach of contract, breakdown in a venture, a government decision affecting a subject) upon circumstances "found" by a judge or a jury; or they might speak *prospectively*, such as when legislation or executive orders seek to guide and declare rules for future behavior.

Ritual accompanies the point of judgment, whatever it be: the moment of sacramental significance. These rituals signify with unmistakable clearness the point at which the blow of judgment is effected: the judge "rules" and "orders" in a formal ritual. Orders are "sealed" by authoritative markings of the court. The jury stands to gives its verdict "so say one so say you all?" they are asked. Statutes are "enacted" when deliberative and voting processes in law-making assemblies culminate in the giving of royal assent. This (in British systems) marks the point of enactment, something once effected by the king actually touching the statute with the scepter. In the days of the death penalty, the judge would place on his head a square of black cloth (a symbol of the black cap of long before). Thus "decorated," the judge would pass sentence with the solemn formulation "you have been convicted of Murder. I therefore pass upon you the sentence of Law. That sentence is that you be taken hence to the place whence you came, that you are there to be hanged by the neck until you are dead, . . . and may God have mercy upon your Soul."[1] Head coverings have deep symbolic value.[2] We glorify princes with the crown, and we cover the head as an act of modesty or humility.[3] We barristers in English and English-style systems wear robes derived from classical dress and we (generally) wear another form of head covering: the wig. Not so long ago, advocates wore the "coif," a close-fitting skull cap or hood.

Law's in-betweeness ritualizes these wondrous Janus-faced moments. This is the point at which the transcendent descends; when the equivocal contracts to the univocal and where the indeterminate meets the determinate. We refer to these points as the "determination," or the verdict (the latter meaning to "speak the truth"). We invite judges in particular (less so legislatures) to "make a determination" about a matter, and in doing so, to bring justice to bear: to enact right against wrong. This determination denotes conclusion; finality and certainty in the act of practical judgment. Justice the overdeterminate calls for this, the determinate. Justice must be *done*. It is not justice until it is done, and it must not be delayed, it must be certain; it must be authoritative by demanding compliance and, if necessary, be enforced and disobedience punished. Doing justice is the corrective for injustice.

The "doing" and the determining is the product of reflection, normally preceded by or accompanying some relational exchange: a trial; parliamentary debate; the making of competing oral submissions or arguments. In courts we have trials and hearings and the judge "reserves" her decision.

1. Bowen-Rowlands, *Judgment of Death*, 239–40.
2. Cairns, *Advocates' Hats, Roman Law and Admission to the Scots Bar; 1580–1812*, 25.
3. Emanations of which we see in the hijab, the yarmulke, and the mantilla.

"*Curia advisare vult,*" "the court wished to be advised" we would say. And "advised" here means to consider, but as part of entering upon a period of reflection directed to transcendent truths ("to catch an echo of the infinite," as we shall see): intellective only in part. This suggests a porosity.

In law-making assemblies (legislatures), there is deliberation, debate, and voting. The politics of these exchanges, and the principle of majority decision (not to be confused with the degeneration into majority *rule*)[4] brings with it, albeit not perhaps in the priestly-like power that attends the judge's function (more on that below), a truly political process and in that way, a wisdom of the multitude, the product, we hope, of striving for the common good, and a melding of contesting forces into the permanent and collective interests of the people.[5]

"*Doing* full justice to what is at play," in the determining process,[6] requires proper attendance to the different stresses revealed by the different modalities of wonder. Bringing the "open infinitude" of the overdeterminate to the pressing determinacy of the moment of judgment without creating a counterfeit infinity is the demanding call that Desmond's chapter makes, and one that presses heavily upon any mediating discipline, and hence for the practices of law and justice.

Enough of the point of judgment.

For a "rule of a moral and logical god above men and even above governments," we seem to wish to substitute the "ideal of a rule of law above men."[7] Law, however, in order that it remain open to its metaphysical sources; to be replenished and corrected by them, must occupy a mediating position and not mistake itself for the *source* of justice. Central to it doing so is to retrieve the life-giving incompleteness, which attends any attempt to know transcendent truths.[8] To accept the incompleteness is to accept this truth, to limit the ambition of law and to understand the role of revelation and finding—and doing—justice.

Our current practices contain more than vestiges or mere echoes of porosity. Our rituals contain an openness to the infinite, by an acceptance of "aboveness" and law as occupying the space between the infinite and doing practical justice.

4. Arendt, *On Revolution*, 16.

5. Bailyn, *To Begin the World Anew*, 118 (citing Alexander Hamilton).

6. Desmond, *The Voiding of Being*, 103 [21–22], emphasis added, and substitution for the infinitive form "to do" the present participle, "doing." The original is "If determinacy is often correlated with univocity, and indeterminacy with equivocity, we need further dialectical and metaxological resources to do full justice to what is at play."

7. Arnold, *The Symbols of Government*, 186; Soper, *The Judge as Priest*, 1210.

8. Pseudo-Dionysius, *The Divine Names*, 52.

Fundamental Incompleteness as Porosity

Oliver O'Donovan's rendering of public authority casts the "ways" of judgment as forbidding human rule to pretend to sovereignty. On this approach, there is only one political society in the end (the New Jerusalem) and there sovereignty is to be found and nowhere else.[9] Law-making therefore, whether it be making law in the form of legislation or rulings by judges is to be conceived within this matrix, and subject to one principal discipline—*to enact right against wrong*.[10] This entails an act of moral discrimination, a pronouncement about a preceding act, and in the practical public context.[11] Sovereignty, in the sense of ultimate law-making authority, and in terms of the higher good of mankind's social destiny, are matters for God's sovereignty and it cannot be devolved to his earthly "viceroys" and vice-regents (as Hobbes and Adam Smith conceived).[12] Only the lower goods of judgment fall to earthly princes.[13]

O'Donovan's frame is one in which earthly political authority is *subordinated*, where, necessarily, its activities are tacit and properly to be within bounds, and even to be distrusted.[14] This concept of limits to political authority by reason of its connection to the overdeterminate is foundational to the exercise of practical judgment. More so this with respect to forward-looking initiative. It is one thing to rule on the past and the present. It is another altogether, given the limitations on our perceptions of the truth, to seek to rule (mainly by legislation) on future behavior.[15] This conception explains why we often have a "disreputable picture of legislative versus an idealised picture of judging."[16] Legislating seeks mostly to govern the future, and judging, the present and the past.

European mediaeval history in its practical formation manifests very much the same ideas. There was a marked absence of any totalizing ambition in the political system. That system showed an inability, or an unwillingness, to be concerned with controlling all forms of social behavior.[17] Power was attributed to the natural world, which was seen as system of primordial

---

9. O'Donovan, *The Ways of Judgment*, 4.

10. O'Donovan, *The Ways of Judgment*, 5.

11. O'Donovan, *The Ways of Judgment*, 7–8.

12. Smith, *The Theory of Moral Sentiments*, III.5.6.

13. O'Donovan, *The Ways of Judgment*, 4.

14. O'Donovan, *The Ways of Judgment*, 5.

15. O'Donovan, *The Ways of Judgment*, 15.

16. Waldron, *The Dignity of Legislation*, 2.

17. Grossi, *A History of European Law*, 1.

rules to be respected.[18] Thus a fundamental *incompleteness* infused political authority. It has only been much later that law has become the expression of a centralized and centralizing will; a legal monism that displaced the pluralism of the Middle Ages.[19] Mediaevals disliked absolutism in the temporal sphere, and conceived only of absolutism in the spiritual realm. This was shown in the temporal's "elaborate distribution of power" and its "sense of corporate life, and its consultation of the various corporate interests through their representatives."[20]

Absolutism is a condition that cannot recognize the given-ness of things, the subordinated nature of political authority, and the sharing of power. The whole concept of one body or person holding a monopoly on political authority over particular territory is very new.[21] Incompleteness of power allowed minorities and diverse legal orders and social groups[22] to survive and flourish, and it allowed room for a whole range of other power-holders: lords, emperors, popes, city states, city leagues, universities, guilds, the manor, local franchises, and myriad social units of organization. Where did they go? They collapsed into an immense and tutelary[23] sovereign state. What we might now see as inevitable, what history tells us is just one possibility. History, however, "covered its tracks well"[24] and we have closed our eyes to other possibilities than the one under which we operate, as if, in some kind of whiggish history, we are standing on the summit. As if all events to this point have led inevitably to this pinnacle.[25]

This movement from a pluralism in political authority to the univocal sovereign state is one that John Milbank has observed in Roman private law's [secular] notion of "dominium" (an unrestricted lordship over what lies within one's power in the household). This was a kind of sheer power. But it came, over time, to be incorporated in the legal *"ius"* or right, once "before and outside the city" and now within it.[26] The *ius*, Milbank conceives, has become detached from what is right or just, to become an active right

18. Grossi, *A History of European Law*, 2.

19. Grossi, *A History of European Law*, 4.

20. Trevelyan, *Some Points of Contrast between Medieval and Modern Civilisation*, esp. 10, 12.

21. Spruyt, *The Sovereign State and Its Competitors*, 1.

22. Grossi, *A History of European Law*, 4.

23. De Tocqueville, *Democracy in America*, Vol. II, ch. VI.

24. Spruyt, *The Sovereign State and Its Competitors*, 1.

25. Butterfield, *The Whig Interpretation of History*, 62–63; Allison, *History in the Law of the Constitution*, 272.

26. Milbank, *Theology and Social Theory*, 13.

over property. This really means that we have come to conceptualize legal rights outside the household with a frame of absolutism or totalizing power: a univocity. The father's power of life and death over his household (*patria potestas*) has been transferred, on Milbank's thinking, to legal rights more generally. We now tend, on this analysis, to see legal rights as domination, rather than sharing in a corporate life, in which the interests in "things" (*res*) might be less totalizing. We could take, as a practical example, rights over land, which, over time, have become more formalized, and have squeezed out rights of occupation held by custom and in common. Clearance of all kinds dispossessed those without dominant, demonstrable (mostly registrable and not community) title[27], yet who depended upon the land for their very survival. Clearance of another kind has also occurred—a formalizing, systematizing power in the form of the modern state—which now has few with which it shares power.

Along these lines we can attend to Desmond's "overdeterminate" modality: we see political authority as exceeding univocal determination, and as having more than one possible emanation. A range of alternative possibilities of arrangement of political authority emerge and we have an opportunity to refresh our practices of law by access to their metaphysical source, something that the positivism of modernity liquidates; a clearance of another kind.

In the fundamental incompleteness that history and theology carve out for us is room for wonder in the true sense; the opposite of its contraction in a closed system, a porosity and a place in which the "happening beyond expectation" can occur, and into the space of the *metaxu*; the "being-betweenness."

All this gives rise to problems for lawyers, trained as we are to eschew unclearness and to advise as accurately as we can, and to know and state all the rules that govern the whole range of human conduct. We have had centuries now of attempts to introduce a scientism of sorts into the frame. Desmond's observations of scientism's tendency to turn itself into the truth of the infinite and to contract the meaning of wonder maps neatly over developments in theories of law. We might briefly survey them to explore Desmond's lower-order modality of wonder; that which tends to orient to the determinate and determinable. I then look to some built-in practices (rather than theories) by which, I suggest, law retains attendance to infinitude.

27. See, for example, in Scotland, Wightman, *The Poor Had No Lawyers*, 49–54, 145, 202.

Legal Science: The Determinate and the Determinable

Scientism made its attempts in law to contract the meaning of wonder; to offer a means of philosophical interpretation of the whole. It arrived with Enlightenment thinking in the eighteenth century and continued in Jeremy Bentham's attempts to codify English law and through to the sociological (instrumentalist) schools of the early twentieth century.

Jeremy Bentham sought "cognoscibility" (knowability) of law by, among other things, its codification. For "cognosciblity" we might confidently substitute univocity and a turn away from the infinite, a kind of turning itself into the truth of the infinite: a self-contained and self-determining system.

Legal science tended to favor legislation over judging. Legislation, as the product of positive enactment and taking the form of textual expression, is more susceptible to systematization and classification and to uniformity. The "science of legislation" took hold in the Enlightenment as part of its self-conscious study of humankind, and tended, as we know, to place humankind's work at the center of things. Incompleteness and imperfection become uncontemplatable on this view; what is not capable of "scientific" expression is not law.

The beginnings were tentative. Early legal "scientists" such as Lord Kames (Henry Home 1696–1782) and Adam Smith (to whom Kames was a mentor)[28] advocated change to judge-made law. One of the principles upon which those reforms were to be based was "utility."[29] This approach took a somewhat traditional view that judge-made law was preferable in method to legislation, but nevertheless formulated a jurisprudence that justified change. The law thus became not an unchanging line of precedent, but a means by which order and justice could be attained in a changing society and to meet such ends as human desires and needs dictated: to "refine gradually as human nature refines."[30]

Adam Smith (1723–90) too participated in the concerted intellectual enterprise of the age to uncover and articulate a science of legislation namely:

> [t]he principles by which those with the power to do so ought to cause changes to law, to enlighten the policy of actual legislators

28. Ross, *Lord Kames and the Scotland of His Day*, 91; Stein, *Law and Society in Scottish Thought*, 159.

29. Lieberman, *The Province of Legislation Determined*, 168, quoting Kames, *Principles of Equity*, 44.

30. Home, *Principles of Equity*, 147.

and encourage them to see what sort of legislative improvements the general interests of the community recommend.[31]

Smith's progress in articulating his science remains less than completely known to us. He finished his *Wealth of Nations* (and published five editions of it in his lifetime). His *General Jurisprudence* (promised by him for decades to be imminent) was never finished. Political economy (the better known of his pursuits), Smith himself said, was one important "branch" of the science of the statesman or legislator.[32] His *General Jurisprudence* was destroyed on his death at his instruction.[33] The work directed at determinacy could not deliver it. Incompleteness remained.

Smith favoured judicial over legislative rule-making: the judge more clearly resembles his impartial spectator; the judge was a reflective autonomous actor; the task of giving effect to the forces at work in law involved indeterminacy, so was much more suited to incremental decision-making than wider scale (i.e., legislative) reform; and the judicial method was better equipped to ascertain and to take account of the kinds of considerations that occupied Smith and governed his natural jurisprudence. Smith distinguished between: "a legislator, whose deliberations ought to be governed by general principles which are always the same" and "that insidious and crafty animal, vulgarly called a statesman or politician."[34]

Smith's science sought to cultivate a natural jurisprudence, and from it emerged principles that "are the subject of a particular science, of all sciences by far the most important, but hitherto, perhaps, the least cultivated, that of natural jurisprudence . . . ."[35]

All this "anticipat[ed] Bentham on utilitarianism."[36] Jeremy Bentham (1748–1832) advocated the use of the principle of utility by the legislature as the touchstone for legislating.

It was no small step from Adam Smith's science to Jeremy Bentham's advocacy of a much more active role for the legislature, and at the expense of the courts. Bentham was the first to assert the "omnicompetence" of the

31. Phillipson, *Adam Smith*, 216 quoting Stewart, *An Account of the Life and Writings of Adam Smith*, 309–10.

32. Smith, *Wealth of Nations*, IV.intro.

33. Winch, "Smith, Adam (bap. 1723, d. 1790)," online ed. There remain, however, student notes of his lectures and from them we gain some insight into his wider thinking: they were discovered only in the 1890s and 1950s, and published as Meek, Raphael, and Stein, *Lectures on Jurisprudence*.

34. Smith, *Wealth of Nations*, IV.ii.39.

35. Smith, *The Theory of Moral Sentiments*, VI.ii.intro.I.

36. Oldham, *From Blackstone to Bentham*, 1657.

legislature.[37] He proclaimed legislation a science. His science, however, was not merely a criticism of the existing body of statute law, but a basis for confident legislator-driven reform, guided by the principle of utility. Nor was his science one that possessed any of the preferences that his near predecessors Kames and Smith had found in the common law's antiquity. And whereas Smith spoke of the "general interests of the community," Bentham's thrust was more reductionistically hedonistic and mathematical; the greatest happiness of the greatest number.

Bentham's approach, including offering utility as a justification to legislate, avoided the limitations that others had identified. He had a disregard for general principles and was impervious to authority.[38] Some have even queried whether Bentham can be regarded as a lawyer. Lawyers, we expect, have an interest in legal technique and recognize some form of legal authority. To Bentham there were only *laws*, made by legislators; an approach that left no room for general principles.[39]

Utility was to become dominant as a justification for the exercise of public authority.[40] It is based upon a conviction that the end of human existence is happiness and that legislation's science is the achievement of laws promoting (aggregate) human happiness.[41] These combined convictions were justifications for legal intervention, and ones that attended not to infinitude, but to a closed human system.

Bentham's science was not without some sense of infinitude. Practically the "most vital part"[42] of it was *laissez faire*. But, this face of limitedness (to quote Desmond) is not to be equated with the fullness of wonder. *Laissez faire* was about the removal of unnecessary restraints placed upon individuals by ancient laws (as altered by haphazard legislation) and also those restraints imposed by positive laws which encroached upon happiness by restricting individual activity. These are pale shadows only of attendance to the indeterminate.

Benthamism was an important catalyst for legislative intervention in every area of life. He legitimized legislative activity. Once merged with Albert Venn Dicey's powerful theorizing of law-as-public-opinion, the twentieth century was set to face the legislative hyperactivity to which we became so accustomed. Is legislative hyperactivity, we might wonder, a

37. Maitland, *A Sketch of English Legal History*, 105.

38. Jolowicz, *Was Bentham a Lawyer?*, 5–7; see also 8 and 12.

39. Jolowicz, *Was Bentham a Lawyer?*, 10.

40. Rosen, "Bentham, Jeremy."

41. Dicey, *Lectures on the Relation between Law and Public Opinion*, 142.

42. Dicey, *Lectures on the Relation between Law and Public Opinion*, 147.

symptom of an inattentiveness to infinitude, an attempt to fill the vacuum
properly left silent, or to be filled by reference to the indeterminate? It
is a space that the determinate often seeks to fill, but never in a way that
satiates and always increasingly hyperactively.[43] "An incessant stream of
lawmaking," O'Donovan offers, has become the "fundamental proof of
political viability."[44] In this there can be a loss of the vital sense of dialogue
between the government and the governed,[45] and thus of the "natural habitat
of truth."[46] The truth that lives in dialogue is squeezed out by contracting the
space for the habitat.

By the late nineteenth century, *laissez-faire* had declined and eventually disappeared as one of the principles by which legislative (in)action
ought to be guided.[47] The principle of utility articulated as part of Bentham's
approach was, as A. V. Dicey himself said, "big with revolution; it involved
the abolition of every office or institution which could not be defended on
the ground of calculable benefit to the public . . . ."[48]

Utility became, without some restraint, or without the rationale of its
unstated restraints being made known, a basis for the rapid and limitless
exercise of political authority. And it fed the view that parliament was supreme—that there is no rule it lacked political authority to make and un-
make, no part of human activity into which it would not inject itself.

The reduction by a scientific approach made the law vulnerable to
instrumentalization. The so-called sociological school (O. W. Holmes Jr.
[1841–1935], Roscoe Pound [1870–1964], and John Dewey [1859–1952]
in particular) illustrated just how vulnerable to powerful interests law could
become. It was a system less interested with law as it should be, but with
law "as it is," and society likewise; in *actual* behavior more than *desirable*
behavior.[49]

Instead of being a system respectful of primordial rules derived from
nature, law was justified in supporting a Darwinian system. Not in creatures
generally, but among humans. Holmes said, as a young man, that the more
powerful interests must be more or less reflected in legislation, which "like
every other device of man or beast [the two are equated here] must tend

43. Virilio, *The Great Accelerator*, 2–3.

44. O'Donovan and O'Donovan, *Bonds of Imperfection*, 213.

45. O'Donovan and O'Donovan, *Bonds of Imperfection*, 220.

46. Pieper, *Abuse of Language, Abuse of Power*, 36.

47. Dicey, *Lectures on the Relation between Law and Public Opinion*, 259.

48. Dicey, *Lectures on the Relation between Law and Public Opinion*, 305.

49. Salmond, *Salmond on Jurisprudence*, 16.

in the long run to aid the survival of the fittest."[50] Legislation, to him, was "necessarily made a means by which a body, having the power, put burdens which are disagreeable to them on the shoulders of everyone else."[51] Thus reduced, law became defined by the followers of this school as what a judge will actually decide: the ultimate in determinacy. Holmes, once said that "civilisation is the process of reducing the infinite to the finite."[52] They sound like Desmondian terms, but the operation here risks reduction rather than attendance of the kind that recognizes a risk or impossibility in the reductive process—the at times necessary incompleteness.

These other famous words from Holmes do suggest an openness to infinitude, even for a jurist considered so distinctively positivist:

> [H]appiness, I am sure from having known many successful men, cannot be won simply by being counsel for great corporations and having an income of fifty thousand dollars. An intellect great enough to win the prize needs other food beside success. The remoter and more general aspects of the law are those which give it universal interest. It is through them that you not only become a great master in your calling, but connect your subject with the universe and catch an echo of the infinite, a glimpse of the unfathomable process, a hint of the universal law.[53]

Dicey (1835–1922) theorized into legal doctrine and practice law as a product of public opinion. His theories saw the rise of parliament now conceived of as sovereign. Without delving into the forces and the significance of how this played out (a topic itself), I would simply observe that the legislature's expropriation of the language of sovereignty migrates the holy to the secular, and transfers care for the holy to the state[54] and strips it of its essence and equivocity. To speak of God as sovereign, and in his absolute right doing all things according to his own pleasure, is quite different from an earthly power claiming such. The word "sovereignty," once detached from a context of a God who (on any view of theodicy) only does good, is power freed from restraint. So to speak of a "sovereign" parliament is to speak of an institution with absolute power to do all things according to its own pleasure (albeit here claiming to act according to the will of electors). And later (only at the end of the last century) came the notion of the

50. Holmes, *The Gas-Stokers' Strike*, 583.

51. Holmes, *The Gas-Stokers' Strike*, 584.

52. Howe, *The Pollock-Holmes Letters*, Vol. 2, 104.

53. Holmes, *The Path of the Law*, 478.

54. Cavanagh, *Migrations of the Holy*, 3, 95–96; Schmitt, *Political Theology*, 36.

sovereign people (that body of people with the formal power to amend a written constitution). Now we even speak of "consumer sovereignty"—that those who consume ought to drive production—by which primacy is afforded to what people hedonistically want. We can thus trace the implosion of the overdeterminate and self-limiting into the banal and unrestrained.

This thinking, although taking us well down the path of a system that offers itself as "having the resources to supply all answers to all the meaningful questions" (as Desmond says), have not succeeded.[55] It has not displaced the need for practices that attend to the infinite.

Attending to the Infinite: Built-in Practices

If, as Desmond points out, the infinite speaks to us through silence not thought, and to test whether law is really a system that can supply all answers from within, we can look to the practices and rituals of law's built-in attention to the infinite, even when acting practically. These silently speaking habits and ingrained structures are more than echoes of past arrangements; they are time-out-of-mind (i.e., the unthought) rituals. In a discipline steeped in tradition, and the need constantly to mediate the tension between it and change, rituals are much deeper than they first appear. I mention just three of them here, but observe that our life as lawyers is filled with conventions, etiquette, habits of mind, rituals, and obediences that attend, in various ways, to the infinite.

Narrative: Unthought Justice

Trials are narrative competitions:

> Each side tells a story, and tries to convince the jury (or judge) to buy his particular vision and version of fact. In an important sense, neither is "true" or "false." The two sides spar with each other before the trier of fact; each embroiders and displays its message with slogans and narrative bits which are thought to be particularly compelling, logical or attractive.[56]

The equivocity of these narratives (competing versions by the various "sides") has to merge in a univocal account of the material facts. The way in which facts are assembled (here we move from fact to *factum*, as something made), the chronology of events; what is to be omitted, what ought

55. Desmond, *The Voiding of Being*, 97 [16].
56. Friedman, *Law, Lawyers and Popular Culture*, 1595.

be emphasized; how should certain things be characterized, what language might be used to describe them are all part of this exercise. One senior judge on a final appellate and constitutional court in Australia has spoken of how the trial judge "has to construct the ultimate narrative, the judgment, and the factual findings, which will determine not only the trial, but also the first and second appeals if they are launched."[57] He speaks of the relation of law to literature, how, drawing on the work of Martha Nussbaum, knowledge of creative literature is "highly desirable" for judicial work, and how it is a major source of "knowledge," and "essential for compassion and tolerance."[58] An acquaintance with literature enables the judge to "deposit the case flexibly in its appropriate department of the common law."[59]

Justice Posner has pointed to the insights judges can obtain from literature that have nothing to do with effective presentation but have to do with "the spirit, meaning of values found in literature, and in a rough sense with content rather than form."[60] He goes on, in a way we might see as an attendance upon infinitude, to say:

> The relevant content, however, is not necessarily or even primarily paraphrasable content, which is the focus of the moralizing tradition in criticism. . . . Moralistic or didactic critics hold with varying degrees of emphasis that the function of literature is to edify and that the canon should be confined to those works (if any—Plato thought that there were few, and Bentham thought there were none) that do edify.

That a judge might, in the midst of the competition for facts, and in "depositing the case flexibly" in the broader legal framework, draw upon edifying narrative is to attend to the infinite, and to admit of porosity.

Juries: Adjudication in Secret Silence

Juries, as we have seen, determine facts. They hear evidence in public and deliberate in private. Unanimity was once enough, and now a majority (albeit a safe one) on occasion may be required. The jury is the "voice of the country," and, as there is only one truth, that voice had to be unanimous. "A

57. Callinan, *The Narrative Compels the Result*, 319.
58. Callinan, *The Narrative Compels the Result*, 324.
59. Callinan, *The Narrative Compels the Result*, 324.
60. Posner, *Law and Literature*, 209.

unanimous verdict . . . which was regarded as representative of the country, an expression of its sense, carried a supernatural weight."[61]

The jury involves placing ordinary people "in the seat of the judge."[62] It divides power in a trial between the judge (who is concerned with questions of law) and the people (to whom is reserved the determination of the facts).

Jury deliberations remain secret. The verdict they pronounce is not to be inquired into, and no reasons are given. The process is not deliberately intellectualized. It is ordinarily an offence for jurors to impart information about their deliberations to anyone other than a juror, and so too for others to procure that information from them. These are solemn obligations.[63]

To have a well-accepted practice for the ascertainment of truth that lies outside a "thought" process, one that need not be externally justified or reasoned or rationalized, is again to permit of porosity, of the infinite speaking in silence.

Judge-Pronounced Law: "living oracles"[64]

The law lies, it was said, in the belly of the judge, *in gremio iudicum*. The judicial function (at least as traditionally cast) was to *discover* or *find*, not to *make* the law: *iudicus est ius dicere non dare* (it is the duty of the judge to declare the law, not to enact it). The declaratory theory of law holds that what judges do is not a source of law, but evidence of what it really is.

In these formulations we find deeply built-in, the notion of the limited judicial function, an acceptance of what is above and beyond, and the need for an attendance not to the judge's own perceptions of justice, but to "given" precepts.

A great deal could be said about the English common law and its porosity. It is not often written about directly. After centuries of promotion of the scientific, the logical, and the rational, the modes by which we maintain (expressly at least) access to metaphysical sources has necessarily contracted. But as a result, they have remained hidden in plain sight within a system that has tended only to see and to value what is posited. "Positive," in a legal setting, is that which denotes methods adopted from the natural sciences, purporting to maintain strict value-neutrality.[65] It was meant to

61. Levy, *The Palladium of Justice*, 43

62. Forsyth, *Trial by Jury*, 357.

63. *Attorney-General v New Statesman and Nation Publishing Co Ltd* [1981] 1 All ER 644 (Widgery CJ).

64. Blackstone, *Commentaries*, i. 68.

65. Pearce, *Science Organized*, 441.

convey a restriction to observed facts, and to separate off speculation as to underlying causes or forces. What cannot be directly observed does not count.

If positivism hold that genuine knowledge is that which we derive from sensory experience, as interpreted through reason and logic, law consists of the commands of human beings. Things with tangible, physical form are easier and simpler to experience, and there is an immediacy and an ease to giving priority to things we can see and feel, and also things we have made ourselves, which, by definition, we must understand, or at least feel, that, as our progeny, we own.

Over time, text and legislation have become more important, and positive enactments of our own creation and within our own control have become frequent. The organic, the spontaneous, the grown, the evolutionary, and common consciousness have all come to be considered less worthy than the things we construct ourselves. Legislation, codes of practice, policies, regulations, rules—and how many other various named enactments—are considered rational and reliable.

But beneath this, we preserve a healthy judge-directed system. It is they who interpret text (including by getting to say what was the intention of parliament in enacting particular legislation), they who decide facts (if no jury), they who rule, and they who "deposit the case."

Conclusion

To the lawyer, the legislator, and the judge, Desmond's modalities keep open an awareness of counterfeits of the indeterminate and our means of attendance to it. The constant demand to "do" practical justice can render those administering justice specially susceptible to a scientism that promises an exhaustive or self-contained system. This counterfeit infinity risks contracting wonder into a mathematically attractive system, with which its authors can marvel at its completeness, its constructedness, and its capacity to allow justice to be dispensed efficiently (i.e., quickly). But without constant attention to the refreshment and correction that the indeterminate supplies, without its equivocity, we are left with univocal, totalizing, absolute power; a power that permits of no "original porosity that is open to being" and susceptible to instrumentalizing co-option.

Bibliography

Allison, John. "History in the Law of the Constitution." *The Journal of Legal History* 28 (2007) 263–82

Arendt, Hannah. *On Revolution*. Great Britain: Faber, 2016.

Arnold, Thurman. *The Symbols of Government*. New York: Harcourt, Brace & World, 1962.

Bailyn, Bernard. *To Begin the World Anew: The Genius and Ambiguities of the American Founders*. New York: Vintage, 2003.

Bentham, Jeremy. *The Works of Jeremy Bentham, Published under the Superintendence of His Executor, John Bowring*, Vol 3. Bristol: Thoemmes, 1995.

Blackstone, William. *Commentaries on the Laws of England*. Oxford: Clarendon, 1765–70.

Bowen-Rowlands, E. *Judgment of Death*. London: Collins, Sons & Co, 1924.

Butterfield, Herbert. *The Whig Interpretation of History*. New York: Norton, 1965.

Cairns, John. "Advocates' Hats, Roman Law and Admission to the Scots Bar; 1580–1812." *Journal of Legal History* 20.2 (1999) 24–61.

Callinan, Ian. "Stories in Advocacy & in Decisions: The Narrative Compels the Result." *Texas Wesleyan Law Review* 12.1 (2005) 319–30.

Cavanagh, William. *Migrations of the Holy: God, State, and the Political Meaning of the Church*. Grand Rapids: Eerdmans, 2011.

Desmond, William. *Voiding of Being*. Washington, DC: Catholic University of America Press, 2020.

Dicey, Albert Venn. *Lectures on the Relation between Law and Public Opinion during the Nineteenth Century*. 2nd ed. Reprint, London: MacMillan, 1920.

Friedman, Lawrence. "Law, Lawyers, and Popular Culture." *The Yale Law Journal* 98.8 (1989) 1579–1606.

Forsyth, William. *Trial by Jury*. Jersey City: Linn, 1994.

Grossi, Paolo. *A History of European Law*. Translated by Laurence Hooper. Hoboken, NJ: Wiley & Sons, 2010.

Holmes, Oliver Wendell. "The Gas-Stokers' Strike." *American Law Review* 7.3 (1873) 582–84.

———. "The Path of the Law." *Harvard Law Review* X (1897) 457–78.

Home, Henry (Lord Kames). *Principles of Equity*. 2nd ed. Indianapolis: Liberty Fund, 1767.

Howe, Mark deWolfe. *The Pollock-Holmes Letters: Correspondence of Sir Frederick Pollock and Mr Justice Holmes 1874–1932*. Cambridge: Cambridge University Press, 1942.

Jolowicz, Herbert. "Was Bentham a Lawyer?" In *Jeremy Bentham and the Law: A Symposium*, edited by George Keeton and Georg Schwarzenberger, 1–19. London: Stevens & Sons, 1948.

Levy, Leonard. *The Palladium of Justice: The Origins of Trial by Jury*. Chicago: Dee, 1999.

Lieberman, David. *The Province of Legislation Determined: Legal Theory in Eighteenth Century Britain*. Cambridge: Cambridge University Press, 1989.

Maitland, Frederic William. *A Sketch of English Legal History*. New York: Putnam's Sons, 1915.

Millbank, John. *Theology & Social Theory: Beyond Secular Reason*. Oxford: Blackwell, 2006.

O'Donovan, Oliver. *The Ways of Judgment*. Grand Rapids: Eerdmans, 2008.

O'Donovan, Oliver, and Joan Lockwood O'Donovan. *Bonds of Imperfection*. Grand Rapids: Eerdmans, 2004.

Oldham, James. "From Blackstone to Bentham: Common Law Versus Legislation in Eighteenth Century Britain." *Michigan Law Review* 89.6 (1991) 1637–60.

Pearce, Trevor. "'Science Organized': Positivism and the Metaphysical Club, 1865–1875." *Journal of the History of Ideas* 76.3 (2015) 441–65.

Phillipson, Nicholas. *Adam Smith: An Enlightened Life*. London: Penguin, 2010.

Pieper, Josef. *Abuse of Language, Abuse of Power*. Translated by Lothar Krauth. San Francisco: Ignatius, 1988.

Posner, Richard. *Law and Literature: A Misunderstood Relation*. Cambridge: Harvard University Press, 1988.

Rosen, F. "Bentham, Jeremy (1748–1832)." Online: *Oxford Dictionary of National Biography*, 2004.

Ross, Ian. *Lord Kames and the Scotland of his Day*. New York: Oxford University Press, 1972.

Salmond, John. *Salmond On Jurisprudence*. 10th ed. London: Sweet & Maxwell, 1947.

Schmitt, Carl. *Political Theology: Four Chapters on the Concept of Sovereignty*. Translated by George Schwab. Chicago: University of Chicago Press, 1985.

Smith, Adam. *The Theory of Moral Sentiments. The Glasgow Edition of the Works and Correspondence of Adam Smith, Vol. 1*. Oxford: Clarendon, 1976.

———. *Lectures on Jurisprudence. The Glasgow Edition of the Works and Correspondence of Adam Smith, Vol. 5*. Oxford: Clarendon, 1979.

———. *An Inquiry into the Nature and Causes of the Wealth of Nations. The Glasgow Edition of the Works and Correspondence of Adam Smith, Vol. 2*. Oxford: Clarendon, 1981.

Soper, Philip. "Metaphors and Models of Law: The Judge as Priest." *Michigan Law Review* 75.5–6 (1977) 1196–1213.

Spruyt, Hendrik. *The Sovereign State and Its Competitors: An Analysis of Systems Change*. Princeton: Princeton University Press, 1996.

Stein, Peter. "Law and Society in Scottish Thought." In *Scotland in the Age of Improvement: Essays in Scottish History in the Eighteenth Century*, edited by Nicholas Phillipson and Rosalind Mitchinson, 148–68. Edinburgh: Edinburgh University Press, 1970.

Stewart, Dugald. "An Account of the Life and Writings of Adam Smith, LLD." In *Essays on Philosophical Subjects*, edited by W. P. D. Wightman, J. C. Bryce, and I. S. Ross, 269–332. New York: Oxford University Press, 1980.

De Tocqueville, Alexis. *Democracy in America*, Vol. II. English ed. Edited by Eduardo Nolla. Indianapolis: Liberty Fund, 2012.

Trevelyan, George Macaulay. "Some Points of Contrast between Medieval and Modern Civilisations." *History* 11.41 (1926) 1–14.

Virilio, Paul. *The Great Accelerator*. Cambridge: Polity, 2010.

Waldron, Jeremy. *The Dignity of Legislation*. Cambridge: Cambridge University Press, 1999.

Wightman, Andy. *The Poor Had No Lawyers: Who Owns Scotland (and How They Got It)*. Edinburgh: Birlinn, 2011.

Winch, Donald. "Smith, Adam (bap. 1723, d. 1790)." Online: *Oxford Dictionary of National Biography*, 2004.

10

Life's Wonder

Simon Oliver

THE LAUDING OF CURIOSITY as an intrinsic feature of modern science and driver of its technological success is central to late modern culture: science unmasks nature's secrets and technology puts nature to work. The curious mind is the living mind, discovering unforeseen possibilities, solving intractable puzzles, and explaining baffling mysteries. Curiosity generates wonder.

Nevertheless, the echo of the ancient warnings against curiosity—its tendency towards power and aimless self-indulgence, which are devoid of respect for the object of study—continue to reverberate. For the Stoics, Augustine, and Thomas Aquinas, curiosity lacks the reverence that attends to the sheer givenness of things. "The desire to know" gives way to "the demand to know." For the ancient tradition, modern thought has the matter the wrong way round: properly ordered enquiry—*studiositas* rather than *curiositas*—must *begin* with wonder.

What are the consequences of the tendency in modern thought to laud curiosity as the beginning of enquiry and sunder it from astonishment at, and reverence towards, the givenness of things? Does the dearth, or even death, of astonishment harbour terrible possibilities? In his meditation on the dearth of astonishment and the limitless curiosity that characterizes modern scientism, explored with characteristic metaphysical profundity, William Desmond suggests so.[1] In this brief response to Desmond's essay, I will describe what I understand to be one of his concerns with wonder

1. Desmond, "The Dearth of Astonishment: On Curiosity, Scientism, and Thinking as Negativity," chapter 3 in *The Voiding of Being*.

139

and its constriction in the scientistic interpretation of science: the loss of science's proper object of study and the pursuit of control rather than truth. The satiation of limitless curiosity becomes its own end, and this often gives way to a violent control of nature. I will then offer some concise historical considerations of curiosity at the birth of modern science in the seventeenth century, pointing to its enthronement of curiosity as a central virtue of the experimental philosopher and *virtuoso*. Although early natural science lacks technological application, its curiosity is already devoid of wonder and reverence towards the sheer givenness of being. In a final section, I will make some tentative suggestions, namely that the loss of the distinction between nature and art, and the animate and inanimate, is the outcome of early modern scientific curiosity, and this has profound consequences: science in its scientistic guise loses its true object of study where "the overdeterminateness of being" is most routinely apparent, namely "life."

Astonishment, Perplexity, Curiosity

In his exploration of the dearth of astonishment and the limitless curiosity that characterizes modern scientism, William Desmond offers a taxonomy of wonder. The first modality of wonder, astonishment, arises with our "porosity to the 'that it is at all' of being."[2] "Astonishment" captures what Desmond calls the "ontological bite" of otherness—an encounter, occasionally fearful and brutal, with the overdetermined abundance of being. Astonishment is a pre-reflexive moment of sheer wonder prior to all reflection and deliberation. This is not, for Desmond, a confrontation with a void of *inde*terminacy—a nothingness—but an encounter with the plenitudinous matrix—the *over*determinate—from which being becomes more determinate.[3]

The second modality of wonder is perplexity, in which we are beset by doubt and uncertainty. We are troubled by the "knots" of the world—the tangled questions it poses and the confounding problems it presents. In perplexity, there is a tense stand-off between our awareness of the "too-muchness" of being encountered in astonishment and our desire to know that overdetermined abundance. We are tempted to reduce the abundance of being to something measurable, yet we remain haunted by the "ever more" of being's gift. In perplexity, there is a blend of the overdeterminate (the sheer abundance of being that is captured in the exclamation "it is!") and the indeterminate (the unspecified and unspecifiable, captured in the exclamation "what is?!"). For Desmond, perplexity can by characterized by

2. Desmond, *The Voiding of Being*, 107 [25].
3. Desmond, *The Voiding of Being*, 103 [21–22].

horror as our rational nature is overwhelmed by the saturated equivocity of being; we struggle to classify, compare, measure, or order. There is both the terrible poverty of not knowing *and* the insatiable desire to know in the face of the unrelenting light of being.

The third modality of wonder, curiosity, is more savvy and less primal than perplexity. Here, we find as univocal a grasp as possible of determinate being, beginning with the question "what is it?" Curiosity seeks to determine the object of its knowledge as a "this" or "that" and to bring the astonishment and perplexity at being's abundance—the "that it is at all"—under the ambit of reason's control and measure. Whereas astonishment is the domain of the metaxological and perplexity the domain of equivocity and strangeness, curiosity seeks a univocal grasp of beings. Curiosity is also marked by self-determination, particularly evident in human beings; we receive ourselves in our relation to determined beings, not from an amorphous well of the indeterminate, but from "the primordial givenness that enables determinacy, that companions our self-determination and yet also exceeds or outlives these."[4]

Whilst these modalities of wonder are distinguishable, they are not, as Desmond makes clear, "epistemic layers"; they are internally related to each other.[5] Astonishment is not attainable *as astonishment* in a pure form, for as soon as we are aware of our astonishment, we become reflexive and receive the porosity of our being, giving way to perplexity and eventually curiosity. Desmond's concern is rather with the relation of the three modalities of wonder. He rightly refuses the Hegelian view that being is unfolded dialectically and negatively *from* curiosity *to* some kind of wonder or astonishment, as if astonishment is something we attain or achieve via curiosity. We begin not with indeterminate negation, but overdetermined plenitude. We receive our porosity to being in the surfeit of its donation; from this, we enter perplexity and curiosity, which are always the outcome of primal astonishment. Curiosity does not produce astonishment; astonishment enfolds curiosity as its origin and end.

Amongst Desmond's concerns with modern scientism's curiosity is its tendency to produce a counterfeit infinity in the guise of an unending and restless inquisitiveness. According to the scientistic interpretation of science, it can be limitlessly interested in everything; it engages in a search for a totalizing univocal account of being. Scientism claims that science has the privilege and, in principle, the capacity to answer all meaningful questions.[6]

4. Desmond, *The Voiding of Being*, 104 [22].
5. Desmond, *The Voiding of Being*, 106 [25].
6. Desmond, *The Voiding of Being*, 97 [16–17].

To question that capacity is arbitrarily to curtail human enquiry and sovereign freedom. The curiosity of science becomes its own end; it is not ordered to the contemplation of its proper object, but to the infinite excitation of ever more curiosity in the hubristic task of vanquishing all ignorance. This is not, however, a curiosity open to the infinite itself; it becomes ever more contracted as it seeks to subsume being (assumed to be nothing more than the phenomenologically given) under its own categories and control. This is a counterfeit double that Desmond terms the "infinite desire to desire without the infinite."[7] The character of scientistic curiosity is also a concern. On the one hand, curiosity can be motivated by love of its object and drawn by its complex beauty. Or it may be motivated by hatred of its object's apparent unwillingness to yield to science's methods and give up its secrets. The "caress" of scientific curiosity can be gentle in coaxing forth nature's secrets, or the "caress" can be aggressive and violent in attempting to subjugate the object of its enquiry as it resists the prurient, stabbing finger of science. This violent, impious, or sacrilegious mode of curiosity concerned the Stoics and Augustine. *Curiositas* establishes idols with a view to prurience and control; its opposite, *studiositas*, is fearfully reverent before its subject conceived in the light of wonder.

For Desmond, the constricted curiosity of scientism becomes particularly evident in the marriage of science and technology. The aspiration towards unhindered self-determination (a kind of anarchic freedom) is expressed in the universal character of science and technology, both in economic reach and in our readiness to turn to them as the solution to all crises, including those they create. The curiosity of science and technology is particularly voracious in the pharmaceutical and military industries, which are "mass formations of the marriage of theoretical and practical curiosity" in which "astonishment and perplexity get in the way of scientistic determinability and superimposing self-determination."[8] The whole of being is determined on the basis of "serviceable disposability," namely as servicing our instrumental desire until it becomes unserviceable and disposable. For Desmond, this impinges on wonder as astonishment and perplexity, for both are beyond mere utility. They open before us the simple givenness of being before and beyond usefulness. We stand before that which cannot be exhausted or encompassed; it simply gives itself continually, regardless of serviceability. Our lack of wonder and perplexity before nature, and our tendency to bring it to heel via scientific-technological curiosity,

---

7. Desmond, *The Voiding of Being*, 121 [38].
8. Desmond, *The Voiding of Being*, 123 [39].

renders it serviceable according to instrumental desires expressed in market economics.

Wonder, Curiosity, and the Beginnings of Science

Has modern science always been married to technology such that its curiosity is reduced to "serviceable disposability" and the mere utility of nature? At first glance, the beginnings of modern experimental and mathematical science in the seventeenth century exhibit a degree of wonder that was not yet allied to technological or industrial utility. The founders of the Royal Society in London in 1660 were pressed to establish the purpose of their natural philosophy or "physico-theology." There was no obvious technological or industrial application of their methods or findings, so they put forward a theological rationale in the form of a defence of Christian belief. In the late 1660s, one of the Society's founders, Thomas Sprat (1635–1713), later bishop of Rochester, was already writing a history of the Society: "I now proceed to the weightiest, and most solemn part of my undertaking; to make a defence of the Royal Society, and this new Experimental Learning, in respect of the Christian Faith."[9] Experimental philosophers were referred to as "priests of nature" and revealers of the wondrous mysteries of God hidden within the natural order. According to Robert Boyle, one of seventeenth-century England's most prominent experimental philosophers, experiments were best performed on Sundays as part of the wider church's pattern of worship.[10] By experimental method, two things were to be discerned and admired: first, the power and intricate craftsmanship of God; secondly, the uses to which creatures could be put. In 1688, Boyle wrote a treatise on the final causes of natural things, by which he meant their utility for human beings. He articulated the desire to understand the utility of nature motivated by curiosity:

> There are not many subjects in the whole compass of Natural Philosophy, that better deserve to be inquired into by Christian philosophers, than that which is discoursed of in the following Essay. For certainly it becomes such men to have curiosity enough to try at least, whether it can be discovered, that there are any knowable final causes, to be considered in the works of nature. Since, if we neglect this inquiry, we live in danger of being ungrateful, in overlooking the uses of things, that may give us just cause of admiring and thanking the author of them, and

9. Sprat, *History of the Royal Society of London* (1667), section XIV, 345.

10. See Shapin and Shaeffer, *Leviathan and the Air-pump*, 319.

of losing the benefits, relating as well to philosophy as piety, that
the knowledge of them may afford us.[11]

Curiosity, therefore, was already confined to the "serviceable disposability" of creatures, the knowledge of which would promote piety. The
God espoused by the early modern natural scientists, whose mind was an
object of curiosity, was a technological craftsman whose secret "know how"
could be uncovered and brought within the ambit of mathematics and experiment. The technological was, after all, already present in seventeenth-
century natural philosophy, albeit a divine technology.

Alongside the tradition of experimental natural philosophers, the
seventeenth century also saw the rise of the *virtuosi* and the transformation of curiosity from an ancient vice that endangered the soul to a modern
virtue that accompanied social status. As William Eamon reports, in the
late sixteenth and early seventeenth centuries the English nobility suffered
a significant decline in favor and status.[12] The usual pursuits of civic service
once resourced by the Crown were no longer available, so bored noblemen
sought other activities that could set them apart from both the vulgar and
common, and the pedantry of medieval scholastics. Amongst their interests were education, not as an intrinsically valuable pursuit of truth born
of wonder or a means to serve the monarchy and state, but as an ornament
betokening wealth and social status. Such education was often concerned
with natural wonders and curiosities—the strange, beguiling, and entertaining. Cabinets of curiosities, particularly notable as a feature of grand tours
to Italy, became increasingly popular. "Curiosity" was not merely a trait of
character or state of mind, but a feature of *things*. The diverting, amusing,
and entertaining appeal of curiosities displayed in cabinets, coupled with
their ability to provoke further curiosity in an open-ended cycle of restless
intrigue, meant that they were prescribed by Robert Burton in his *Anatomy
of Melancholy* (1621) as a suitable treatment for melancholy as well as a sign
of prosperity and tokens of status.[13] It was not, however, merely natural
curiosities and the oddities of nature that were of interest to the *virtuosi,*
but also "mathematical magic"—an array of mechanical gadgets and en-
gines, tricks, optical illusions, mathematical puzzles, and the like.[14] These
diversions and amusements involved familiarization with instruments such
as the thermometer and optical lenses that would later become central to
scientific experimentation and the furtherance of curiosity. The instrument

11. Boyle, *A Disquisition about the Final Causes of Natural Things* (1688), A2.

12. Eamon, *Science and the Secrets of Nature,* 302–3.

13. Eamon, *Science and the Secrets of Nature,* 304–6.

14. Eamon, *Science and the Secrets of Nature,* 306.

with perhaps the greatest power to intrigue and mystify was Robert Boyle's famed air-pump which became the centerpiece of the Royal Society at its foundation. It could produce what nature apparently abhorred, namely a vacuum, and thereby exhibited the greatest technological control. It provoked curiosity by producing literally "nothing"—a void space that could nevertheless extinguish the light of a candle and the life of a bird.[15]

The popularity of early scientific instruments lay in their ability to aid the curious in their pursuit of nature's secrets. The outcomes of such experiments—the suffocation of a bird in an air-pump, for example—are not natural but artificial. As Eamon suggests, facts of nature in the seventeenth century were increasingly the products of instruments and machines; modern scientific truth is, from the outset, technologically produced. Telescopes and microscopes brought the invisible under the human gaze, whereas thermometers and barometers measured subtle meteorological phenomena that were otherwise imperceptible.[16] In later laboratory science, discoveries are artificially produced in a strictly controlled environment by being coaxed from nature. According to Eamon, the acceptance of experimental truths in early modern science, manufactured by human art, depended on the view that there is no fundamental distinction between art and nature, for what is produced artificially *is* nature. Francis Bacon, the father of modern scientific method, explicitly rejected the distinction: "Whereas men ought on the contrary to be surely persuaded of this; that the artificial does not differ from the natural in form or essence, but only in the efficient. . . . Nor matters it, provided things are put in the way to produce an effect, whether it be done by human means or otherwise."[17]

However, it is not simply a case of the rejection of the distinction between nature and art; in early modern science nature came to be seen as a work of divine artifice or "technology." Aristotle prioritized nature as that which has an intrinsic principle of motion and rest expressed particularly through a creature's form in such a way that a natural creature's *telos* was intrinsic to its nature—acorns become oak trees and cygnets become swans. The artificial possesses no intrinsic principle of motion and rest—timber does not by nature become a cabinet or a chair but must be fashioned by a craftsman via the extrinsic application of form and *telos*. By the late seventeenth century, all nature was regarded as "artificial" in the sense that it was

15. On the significance of Boyle's air-pump in early modern science, see Shapin and Shaeffer, *Leviathan and the Air-pump*.

16. Eamon, *Science and the Secrets of Nature*, 310.

17. Bacon, "Of the Dignity and Advancement of Learning" (*De augmentis scientiarum*) book 2, in *The Works of Francis Bacon, Vol. 4*, 294–95, quoted in Eamon, *Science and the Secrets of Nature*, 310.

composed of inert matter possessing no intrinsic form. It was governed by a divine artificer via the mathematically expressible laws of nature, hence the modern "design argument" for God's existence was born. Causation was reduced from an intricate blend of Aristotle's four modes of formal, final, material ,and efficient, to the merely efficient. As such, nature was conceived as matter in local motion and, in principle, all its secrets were reducible to, and discoverable by, mathematical physics. There was nothing *intrinsically* different about the operation of mechanical printing press and the human body; the latter was simply vastly more complex and less ready to give up its secrets to the *curiosi*.

The narrowing of causation to the efficient and mechanistic along with a materialist ontology represents a reduction to the univocal in Desmond's sense. All causes are univocal efficient causes, all beings are univocal material beings lacking *intrinsic* purpose and value. Insofar as they lack intrinsic ends, they are technologically manipulable to *any* end. Whilst perplexity arises in the contemplation of being, the method for its overcoming—mathematics and experiment—is plain. Although the curiosities and experiments of early modern science could begin with a general wonderment, the purpose of science was to dissolve wonder by explaining the causes of things in purely immanent mechanistic terms before moving to the next marvel and curiosity in an endless pursuit of nothing in particular, save the establishment of the use of things. What were the consequences of the reduction of nature to artifice and a univocal materialist ontology?

The End of Life

What was lost at the birth of modern natural science was not simply the distinction between the natural and the artificial, but also the associated distinction between the animate and inanimate. For Aristotle, metaphysics is concerned with being *qua* being and natural philosophy is concerned with varieties of motion and particularly self-moving being—namely, life. However, if nature is reduced to mere artifice and creatures are simply matter in different configurations, there is no fundamental distinction between the functioning of a living body and the functioning of a machine. A machine crafted by human hands was not understood as imitating a living creature; machine and organism were understood univocally as various configurations of matter. Dead and inert matter becomes ontologically primary; all wonder can be dispelled in the reduction of living beings to mechanistic processes. Insofar as nature is animated, its source is God who is conceived as a univocal mover of the cosmos, replenishing its motion from time to time as the cosmic clockwork mechanism winds down.[18]

The philosopher Hans Jonas refers to the radical shift in imagination in the sixteenth and seventeenth centuries that saw life reduced to a curious aberration in a tiny and insignificant part of an otherwise material and dead universe.[19] In ancient thought the cosmos was regarded as a living creature, as in Plato's *Timaeus*, or life was thought to be the purpose of the cosmos and therefore philosophy's central concern, as in Aristotle. Early modern natural philosophers gazed into the heavens and saw nothing but inert matter in various orbits. They also gazed inside the human body and saw mere materiality with the consequent loss of a normative conception of human nature.[20] Whereas the fundamental focus of ancient thought was the mystery of life in which death was a problem to be explained, for modern science the universe is fundamentally dead and *life* becomes the problem to be explained:

> Accordingly, it is the existence of life within a mechanical universe which now calls for an explanation, and explanation has to be in terms of the lifeless. Left over as a borderline case in the homogenous physical world-view, life has to be accounted for by the terms of that view.[21]

18. This raises the significant problem of agency within nature. Jessica Riskin remarks, "I think that biologists' figures of speech reflect a deeply hidden yet abiding quandary created by the seventeenth-century banishment of agency from nature: do the order and action in the natural world originate inside or outside? Either answer raises big problems. Saying "inside" violates the ban on ascriptions of agency to natural phenomena such as cells or molecules, and so risks sounding mystical and magical. Saying "outside" assumes a supernatural source of nature's order, and so violates another scientific principle, the principle of naturalism." Jessica Riskin, *The Restless Clock*, 6.

19. Jonas, *The Phenomenon of Life*, 7ff.

20. For Jonas, 1543 is therefore a symbolic year in the emergence of modern science: Nicolaus Copernicus published *On the Revolutions of the Celestial Orbs* and Andreas Vaselius published *On the Fabric of the Human Body*. See Jonas, *Philosophical Essays*, 52–59.

21. Jonas, *The Phenomenon of Life*, 10. In this context, it is really no surprise that scientism can be indifferent as between the pharmaceutical industry (ostensibly the protector of life) and the military industry (the engineer of death). They are both aspects of a single mode of enquiry that, because of its ontological materialism, sees no fundamental difference between the animate and inanimate, the living and the dead. It might be objected that the use of the term "dead" in this context disingenuous when what is meant is the mere indifference of matter. "Dead" has an antithetical meaning in relation to that which was once alive. Jonas deals this issue in *The Phenomenon of Life*, 12.

The materialist ontology that characterized seventeenth- and eighteenth-century natural philosophy meant that physics and chemistry came to be seen as the fundamental or "hard" sciences. Perplexity could be dissolved and curiosity satiated by reducing all phenomena to their material constituents. When biology arose as a distinct science of life in the nineteenth century, it revived concepts of agency, purpose, concern, and value that seem to be features of living organisms, although these were thought to be metaphorical or heuristic. It was assumed that biology would eventually be reduced to chemistry, then to physics. Insofar as science is reducible to physics, it is confined to the univocal. Its processes are, in principle, mathematically expressible and predictable, and therefore manipulable. Of course, the reducibility of nature to classical mechanics is no longer tenable in the wake of twentieth-century quantum physics and thermodynamics. Moreover, as scientism's materialist orthodoxy has hardened, so its grasp of what matter *is* has waned. Perplexity returns: scientism might insist that matter is all there is, but this claim amounts to very little if matter is a mystery.[22]

Could it be that, as modern thought inverted curiosity and astonishment, seeing the latter as an occasional but diminishing by-product of the former rather than its fount, so early modern natural philosophy inverted true science, seeing life as an occasional but doomed by-product of matter? Might one regard biology as the more fundamental science that resists reduction to chemistry and physics, to matter in its different configurations? Whilst life is notoriously difficult to define, its identification with metabolism marks its irreducibility to mere materiality. Metabolism is the set of chemical reactions serving three purposes: the anabolic conversion of nutrition into the building blocks of cells; the catabolic breaking down of nutrition; and the elimination of waste. The metabolic process is replete with metaphysical significance, according to Hans Jonas: "The exchange of matter with the environment is not a peripheral activity engaged in by a persistent core: it is the total mode of continuity (self-continuation) of the subject of life itself."[23] The living organism holds itself back from entropic equilibrium and constantly exchanges matter with its environment. Unlike the source of energy for a machine (electricity for a photocopier, fuel for a car), the matter assimilated by the organism in the form of food becomes part of the organism. The living organism persists as its same self despite never being the same matter; the whole, its living form, precedes and defines

22. On the correlation between metaphysical materialism and the unfolding mystery of matter in physics, see Oliver, "Physics without *Physis*."

23. Jonas, *The Phenomenon of Life*, 76 n.13.

its material nature. The living organism is an irreducible unity that cannot be explained in material mechanistic terms.

Because the living organism holds itself and displays concern for itself, it possesses freedom from merely material processes. It also holds within its form a teleological structure and therefore purpose and value beyond "serviceable disposability." Here, in the study of life, science blends with the metaphysics of purpose and value. Could it be that science, when life is its primary object of study, is more conducive to perplexity and recalls a primordial state of astonishment, which is the basis of true and reverent enquiry? We can recall that, in Desmond's taxonomy of wonder, perplexity is the domain of the equivocal: we struggle to classify, compare, measure, or order. Life is equivocal in the sense that it is infinitely various, from amoeba to human consciousness, and prompts in us the greatest degree of perplexed wonder. Yet life is also elusively analogical, for all life is intertwined; the single concept of "life" is never applicable in purely univocal terms. Whilst life gives itself to be known in the self-moving organism, it also contains an irreducible interiority that is not amendable to mere curiosity but is a source of wonder and perplexity. This is most evident in human conscious life—the "what it is to be me" that remains inaccessible to others, yet a mystery that concerns others. We are perhaps most astonished at the life that comes from our own life, namely our children. The sheer givenness of a child's life, in excess of all material process and human artifice or will, and the pre-reflexive astonishment at our first encounter with our own children, reminds us of life's metaphysical significance and a child's primordial astonishment. Finally, life always returns, forever repeated in non-identical fashion as spring emerges from winter, and therefore becomes a sacrament of the agapeic character of being's primal gift.

The collapse of the distinction between nature and art, and the animate and the inanimate, at the advent of modern science in the sixteenth and seventeenth centuries meant that "mere curiosity"—a kind of prurient and quizzical gaze—could be the engine of enquiry. Yet such a gaze could not but destroy its object because it found therein no *intrinsic* self-concern, purpose, or value. Early modern science could not be studiously reverent because there was nothing to reverence. The indeterminate was rendered determinate under the power of a metaphysical materialism. The traditional hierarchy in which astonishment was the fount of studious enquiry was reversed: curiosity could be the occasional source of fleeting and fickle wonder. A broadening of science's metaphysical horizons that reinstates the distinction between nature and artifice, the animate and inanimate, may also reinstate "life" as its proper object of enquiry, understood analogically as the object of wonder in its three modes, and the sacrament of the agapeic origin of overdeterminate being itself.

Bibliography

Bacon, Francis. *The Works of Francis Bacon: Volume 4: Translations of the Philosophical Works 1*. Edited by J. Spedding, R. Ellis, and D. Heath. Cambridge: Cambridge University Press, 2011.

Boyle, Robert. *A Disquisition about the Final Causes of Natural Things*. London, 1688.

Desmond, William. *The Voiding of Being: The Doing and Undoing of Metaphysics in Modernity*. Washington, DC: The Catholic University Press of America, 2019.

Eamon, William. *Science and the Secrets of Nature: Books of Secrets in Medieval and Early Modern Culture*. Princeton: Princeton University Press, 1994.

Jonas, Hans. *The Phenomenon of Life: Towards a Philosophical Biology*. Evanston, IL: Northwestern University Press, 2001.

———. *Philosophical Essays: From Ancient Creed to Technological Man*. New York: Antropos, 2010.

Oliver, Simon. "Physics without *Physis*: On Form and Teleology in Modern Science." *Communio: International Catholic Review* 46.3–4 (2019) 442–69.

Riskin, Jessica. *The Restless Clock: A History of the Centuries-Long Argument over What Makes Living Things Tick*. Chicago: University of Chicago Press, 2016.

Shapin, Steven, and Simon Shaeffer. *Leviathan and the Air-pump: Hobbes, Boyle, and the Experimental Life*. Princeton: Princeton University Press, 2011.

Sprat, Thomas. *History of the Royal Society of London for the Improving of Natural Knowledge*. London, 1667.

11

Being in Control

Michael Hanby

WILLIAM DESMOND IS A prime number, singular and irreducible. This is evident in the very language with which he philosophizes, language that somehow manages to locate him squarely and recognizably in the great tradition of questioning extending from Plato to Heidegger, while remaining as particular as the Irish poets who haunt his thinking as muses. Indeed, there is something inherently poetic about Desmond's philosophy—which I mean as praise and not as a criticism—as if it has been visited by spirits bearing a vision that overflows the words employed to contain it. This endows his thought with that peculiar quality that the novelist Flannery O'Connor ascribes to the genre of story. "A story," O'Connor wrote, "is a way to say something that can't be said in any other way, and it takes every word in the story to say what the meaning is."[1]

This leaves the would-be interpreter of Desmond with something of a dilemma. To attempt to translate Desmond's vision from the particular language of its expression into a more familiar idiom is inevitably to omit something essential that can only be apprehended and expressed—that only comes to view—in the inimitable language of the original. On the other hand, to attempt to think in Desmond's singular idiom is to risk producing a counterfeit double of his thought.

It is in full consciousness of this dilemma, and of my own limitations of vision and expression, that I would like to reflect on two aspects of Desmond's essay in this volume, with whose thesis I am in fundamental

---

1. O'Connor, *Mystery and Manners*, 96.

151

agreement. I wish to reflect on what one might call, using or modifying Desmond's own lexicon, the different *modes* of determinacy under which beings manifest themselves to us, both to further break open the possibility that there might be modes of apprehension, many of which we may be in danger of losing, that do not negate the original overdeterminacy—the "too-muchness" of being that generates its own porosity in us at the origin of thought, but also to question the adequacy of the distinction between science and scientism with which Desmond brackets his essay. Desmond seeks to "reflect on the temptation to contract the meaning of wonder in the scientistic interpretation of scientific curiosity." To refer to this contraction as a "temptation" is to suggest the possibility of a science that does not succumb to it, that is, to a distinction between science (in its modern sense) and scientism. In confirmation of Desmond's concern and the profundity of his diagnosis, but perhaps going beyond them, I wish to suggest that the very mode of modern scientific cognition is so constituted by its having always already succumbed to this temptation that if we wish to sustain a distinction between science and scientism, it would require us to reconceive of the nature and practice of science itself in a way that few of us are prepared, much less able, to do.[2]

Desmond takes up scientific curiosity as one of the three modes of wonder, the primitive openness generated in us by being's self-communication and its infinite excess to that communication.[3] Whereas the "desire

2. I am reminded here of C. S. Lewis's remark near the conclusion of *The Abolition of Man*. "Is it, then, possible to imagine a new Natural Philosophy, continually conscious that 'the natural object' produced by analysis and abstraction is not reality but only a view, and always correcting the abstraction? I hardly know what I am asking for. . . . The regenerate science which I have in mind would not do even to minerals and vegetables what modern science threatens to do to man himself. When it explained it would not explain away. When it spoke of the parts it would remember the whole. While studying the *It* it would not lose what Martin Buber calls the *Thou*-situation. The analogy between the *Tao* of Man and the instincts of an animal species would not mean for it new light cast on the unknown thing, Instinct, by the only known reality of conscience and not a reduction of conscience to the category of Instinct. Its followers would not be free with the words *only* and *merely*. In a word, it would conquer Nature without being at the same time conquered by her and buy knowledge at a lower cost than that of life." Lewis, *The Abolition of Man*, 79.

3. There is a family resemblance between Desmond's differentiation of the modalities of wonder, as well as the dangers inherent in these modes, and Hans Urs von Balthasar's distinction between wonder and admiration. "Without doubt the phenomenal world contains on all sides an objective order which is not imposed by man, and thus a beauty, the legitimacy of the premise is repeatedly confirmed for him that there is within Nature a greater objective ordering of things than he had previously recognized. Every theoretical science with a practical application, such as medicine

to know" is derivative of this original wonder, scientific curiosity is derivative of this desire. The first mode, astonishment, corresponds to the
"too muchness" of being's over-determination, what I sometimes call its
intensive infinity.[4] The second mode, perplexity, corresponds to the apparent indeterminacy of being, which elicits the desire to know on our part.
Curiosity, then, corresponds to what is determined or determinable and
is therefore a limiting mode of the original porosity. This is the ground of
Desmond's worry that scientific curiosity is easily seduced into thinking of
itself as "self-born, self-activating"—forgetting its ontological and historical
conditions of possibility—and thereby takes on "an energy of determination
that rides over the original givenness of this porosity." This forgetfulness
corresponds to science's "lack of thinking" diagnosed so provocatively by
Heidegger, and it sounds, initially, like a problem that could be remedied
by a kind of philosophical mindfulness. But Desmond is clear: "The lack of
thinking is not remedied by the supplement of self-reflection. The mystery
of being in its given otherness is also at stake."

But, of course, not all forms of determination are created equal. My
determination of the sun as I sit on a mountainside at daybreak contemplat-
ing its luminosity as a visible image of an intelligible light is different not
just in degree but in kind from my capture of the sun's light for the purposes
of converting it to electricity. The former determination does not negate
the "wonder before the being there of being and beings." It augments it, so
to speak, precisely by allowing the "too-muchness" of being and beings to
come to a self-exceeding expression. The sun under such a determination
is not just a "giant ball of gas," a source of energy, or—as an astrophysicist

---

or physics, lives from this perennial assumption which forever proves itself anew. So
much so, that on this basis philosophy dares to make an ultimate leap forward by projecting a totality of sense upon the totality of the actuality of Being in such a way that
now necessity is predicated of the latter. Then Being becomes identical with the necessity to be, and when this identity has been taken up by reason, there is no longer any
space for wonder at the fact that there is something rather than nothing, but at most
only for admiration that everything appears so wonderfully and 'beautifully' ordered
within the necessity of Being." Balthasar, *Glory of the Lord V*, 613.

4. While this "intensive infinity" can be registered phenomenologically apart from
any explicit theological commitments, it is also an ontological implication of a theology of creation, where 'creation' is not only a designation of origin but of the *ex
nihilo* structure of being, which imbues each thing and all things with the character of
unrepeatability and interiority. That a creature is constituted in its very being by the
interior presence of the bottomless God, who as the giver of that being presupposed in
every subsequent determination is "In all things, and innermostly" (Aquinas, *ST*, I.8,1,
resp.), means that a certain "bottomlessness" extends by analogy to the things of the
world and that a certain *via negativa*, whose primitive form is adoration rather than
domination, must belong to any adequation between the mind and being.

once said to me—"a factory for making chemicals," it is a visible symbol of a reality in which it partakes but which exceeds it. As a symbol, it represents something more and beyond; it is replete with a meaning that it cannot itself contain. Implicit in this mode of determination—operative in it, we might say—is an entire ontology that does not foreclose on being. Pushed to its ultimate implications, which is another way of saying by remaining a question open to the "more," it takes us to the brink of "coming-to-be" in its distinction from "becoming."

Scientific inquiry commences from within this world of meaning, a whole comprised of intelligible wholes. This is true both on the side of the object—its unity providing the tacit ground of intelligibility for all subsequent abstractions, both experimental and mathematical[5]—and on the side of the subject, whose existence is characterized by the inseparable union of "inwardness and outwardness in one" and whose every thought and action therefore recapitulates the entire history of being that is its condition of possibility.[6] Scientific cognition therefore never quite succeeds in attaining what Hannah Arendt called the "universal" and "astrophysical" perspective, the Archimedean point outside of nature fantasized by Descartes.[7] Descartes

5. "For in order that we may formalize the relations that constitute a comprehensive entity, for example, the relations that constitute a frog, this entity, i.e., the frog, must be first identified by tacit knowing, and, indeed, the meaning of a mathematical theory of a frog lies in its continued bearing on the tacitly known frog." Polanyi, *The Tacit Dimension*, 20–21.

6. "All that precedes only expresses the inevitable exigencies of thought and practice. That is why it is a system of scientific relations before appearing as a chain of real truths. In thinking and in acting, we imply this immense organism of necessary relations. To lay them out before reflection is simply to unveil what we cannot help admitting in order to think, and affirming in order to act. Without always noting it distinctly, always we are inevitably brought to conceive the idea of objective existence, to posit the reality of objects conceived and ends sought, to suppose the conditions required for this reality to subsist. For, not being able to do as if it were not, we cannot include in our action the indispensable conditions for it to be. And reciprocally, what cannot be immanent in thought, we cannot tend to make immanent to ourselves through practice. The circle is closed." Blondel, *L'Action*, 421.

7. See Hannah Arendt: "For whatever we do today in physics—whether we release energy processes that ordinarily go on only in the sun, or attempt to initiate in a test tube the processes of cosmic evolution, or penetrate with the help of telescopes the cosmic space of two and even six billion light years, or build machines for the production and control of energies unknown in the household of earthly nature, or attain speeds in atomic accelerators which approach the speed of light, or produce elements not to be found in nature, or disperse radioactive particles, created by us through the use of cosmic radiation, on the earth—we always handle nature from a point in the universe outside the earth. Without actually standing where Archimedes wished to stand (*dos moi pou sto*), still bound to the earth through the human condition, we

himself recognized that this is virtually impossible, which is why he had to invent the fictitious *deus malignus* to negate the otherwise indubitable world of wholes, forms, and qualities, and even then he failed to touch the language in which his pretended doubt was expressed.[8]

Nevertheless, scientific cognition is premised upon an attempted negation of this intelligible world and the "too-muchness" that characterizes both its intensive and extensive infinity. Its founding gesture, repeated in different forms by its philosophical architects in the seventeenth century, is a "no" to being in the fullness of its self-presentation, which includes the sense in which we invariably apprehend it that furnishes science with its own condition of possibility. Rather, science seeks, as Francis Bacon put it, to "take experience apart and analyze it," a process that leaves no room for a wholeness that ontologically precedes its component parts, no room for an interior depth of being that is not a concatenation of surfaces, and no room for the cognitive act by which it is apprehended.[9] Entailed in the structure of scientific cognition, in other words, is what Hans Jonas called a "primary ontological reduction" that leaves the reducer in permanent repeat to the fictitious Archimedean point outside of being in the moment of his reductive theorizing.[10] But reduction to what?

---

have found a way to act on the earth and within terrestrial nature as though we dispose of it from outside, from the Archimedean point. And even at the risk of endangering the natural life process we expose the earth to universal, cosmic forces in nature's household." "In the experiment," she continues, "man realized his newly won freedom from the shackles of earth-bound experience; instead of observing natural phenomena as they were given to him, that is, he placed nature under the conditions of his own mind, that is, under conditions from a universal, astrophysical viewpoint, a cosmic standpoint outside nature itself. Arendt, *The Human Condition*, 262, 265.

8. "I will now shut my eyes, stop my ears, and withdraw all my senses. I will eliminate from my thoughts all images of bodily things, or rather, *since this is hardly possible*, I will regard all such images as vacuous, false, and worthless." Rene Descartes, "Meditations on First Philosophy," in *The Philosophical Writings of Descartes*, Vol. II, 24, emphasis mine.

9. Francis Bacon, "Plan of the Great Renewal," in *The New Organon*, 17. For an expanded critique of Baconian analysis and its accompanying metaphysics see Hanby, *No God, No Science?*, 107–49.

10. To the first point, the inherent reductionism of modern science, Jonas writes, "The key term here is 'analysis'. Analysis has been the distinctive feature of physical inquiry since the seventeenth century: analysis of working nature into the simplest dynamic factors. These factors are framed in such identical quantitative terms as can be entered, combined, and transformed in equations. The analytical method thus implies a primary ontological reduction of nature, and this precedes the mathematics or other symbolism in its application of natures." To the second point about the permanent self-exemption, he says, "It is this reflection that the cybernetician fails to perform.

The key lies in the meaning of the Baconian conflation of knowledge and power, which is decisive for the whole scientific enterprise.[11] This conflation lies at the heart of scientific cognition as a form of abstraction from the paradoxical plenitude of the real, at once simultaneous and successive, which inverts the priority of act and potency and premises the real order upon the artificial one abstracted from it.[12] The forgetfulness of being that Desmond laments is thus an endemic feature of the analytic method. "The chief thing lost sight of in an exclusive application of analysis," Goethe wrote, "is that all analysis presupposes synthesis."[13] The conflation of knowledge and power is not merely a question of motive; this subjectivist misunderstanding is the risk Desmond incurs in speaking in terms of wonder when in fact he is designating an ontological, and not merely epistemological porosity. Nor is this conflation merely a means-end relationship, knowledge *for the sake of* power. Modern science does not simply know natural phenomena for the sake of controlling them, in order to put theory to later practical uses, though of course this is also true. Rather, science knows natural phenomena *by means of* controlling them, which implies a much more intimate co-penetration of *theoria* and *praxis*, knowing and making, at the very outset. This is what

He himself does not come under the terms of his doctrine. He considers behavior, except his own; purposiveness, except his own; thinking, except his own. He views from without, withholding from his objects the privileges of his own reflective position." He continues, "The attempt, therefore, in disowning itself as evidence of its subject matter, contradicts itself with the kind of understanding it achieves of its subject matter. In eliminating itself from its account, it makes the account incomplete, yet does not tolerate a completion that would transcend the self-sufficiency of its principle, in virtue of which the account is closed in itself. Thus the attempt not only leaves itself unaccounted for, and unintelligible in its own terms, even more, with the epiphenomenal depreciation of inwardness, it invalidates its own finding by denying to thinking a basis of possible validity in an entity already completely determined in terms of the thoughtless. It is the Cretan declaring all Cretans to be liars." Jonas, *The Phenomenon of Life*, 200, 123–24, 134.

    11. This is why John Dewey calls Francis Bacon the "real founder of modern thought." Dewey, *Reconstruction in Philosophy*, 28.

    12. For more on the nature of scientific abstraction, see Hanby, *No God, No Science?*, 107–49, 375–415 and Schindler, *Ordering Love*, 383–429. This paradoxical plenitude has been recognized in various ways; the language of simultaneity and succession is from Goethe. It underlies his morphological investigations and his development of concept of the *Urpflanze*, as well as the priority of analysis to synthesis. The later we see mirrored in the Aristotelian-Thomist priority of *intellectus* to *ratio* as the latter's origin and end. See Goethe, *Goethe's Botanical Writings*, 219. See Aquinas, *In Boeth. De Trin.*, vi.1, ad 4.

    13. Goethe, *Goethe's Botanical Writings*, 239.

Bacon meant in proposing to "conquer nature by work."[14] The "truth" of scientific knowledge—which had to be fought for in the face of the old contemplative ideal and its equation of truth and being—is precisely identical to the kinds of control we are able to exercise over natural phenomena: in the successful replication of experiments, the prediction and retro-diction of events, or our capacity to "generate and superinduce on bodies a new nature or new natures" and to effect "the transformation of concrete bodies from one thing into another within the bounds of the *Possible*," descriptions that define the task of the new sciences and establish the boundary that they will perpetually transgress.[15] Truth in this sense is not being's self-revelation of its given actuality, but its response to a prior intervention on our part that determines and measures that revelation under the form of possibility, or, as Bacon says, *power*. Here we are close to Heidegger's "standing reserve," a revealing of the possibilities of being—or being as possibility—as elicited by a "challenging forth."[16]

Desmond is absolutely correct that "the senses of being are at stake" in this apprehension. Scientific cognition determines being in the mode of power or possibility, or more precisely, *our* power to actualize certain policies latent within it under conditions we impose. The sense of being manifest in experimental science is therefore not being in its "in-itselfness" as nature, but being "for us" as artifice, and therefore being as subject to our unmaking and remaking. Desmond's image of a "the marriage of science and technology" is thus misleading inasmuch as it suggests the union of partners already whole and complete unto themselves. The fusion of *techne* and *logos*, making and knowing, is present in science *ab initio*—it is its form and essence. Technology, as Hans Jonas puts it, is the metaphysics of science "come out into the open."[17] And as metaphysics in action, it is the ontology of the age.

Desmond is therefore also right to worry about "our violent intrusion into everything, with no reverence that lets things be as they are, that loves things even as they are, that blights the goodness of creation. He is right to worry about an approach that "does not leave the space of the between open. It does not allow anything to be itself and to reveal itself in its own more intimate ontological terms." Whether he means to echo Galileo's "rape of the senses" in calling such an approach "the rape of an erotic assault" I cannot say, but the objective presence of that echo suggests that this closure is not

14. Bacon, "Plan of the Work," in *New Organon*, 16.

15. Bacon, *New Organon*, II.1.

16. Heidegger, "The Question Concerning Technology," in *Basic Writings*, 321–22.

17. Jonas, *Philosophical Essays*, 48–49.

a second-order, ideological perversion of the modern scientific enterprise. This is the assumption that lies behind the conventional distinction between science and scientism, that the empirical and experimental methods of science are ontologically neutral precisely *as method*.[18] I rather doubt this is what Desmond means in distinguishing between science and scientism, but if science is a kind of metaphysics in act, and if this reductive closure is endemic to that metaphysics, then the distinction between science and scientism is finally untenable. This is why but for a few eccentric exceptions—Goethe and Hans Driesch come to mind—science has successfully resisted successful integration beyond the attitudinal or subjective level in a more comprehensive order of being, reason, nature, and truth. The metaphysics that the sciences put into action through "the project of seamless self-determination" have thus far negated their condition of possibility. Desmond is right to say that "the project to succeed is thus already a failure." But if we wish to endure this failure or even somehow overcome it, we must also see that this failure is the key to science's overwhelming success.

Bibliography

Arendt, Hannah. *The Human Condition.* 2nd ed. Chicago: University of Chicago Press, 1998.

Bacon, Francis. *The New Organon.* Edited by Lisa Jardine and Michael Silverthorne. Cambridge: Cambridge University Press, 2000.

Balthasar, Hans Urs von. *Glory of the Lord, Vol. V: The Realm of Metaphysics in the Modern Age.* Translated by O. Davies, A. Louth, B. McNeil, J. Saward, and R. Williams. San Francisco: Ignatius, 1991.

Blondel, Maurice. *L'Action: Essay on a Critique of Life and a Science of Practice.* Translated by O. Blanchette. Notre Dame, IN: Notre Dame University Press, 1984.

Descartes, Rene. *The Philosophical Writings of Descartes*, Vol. II. Translated by J. Cottingham, Robert Stoothoff, Dugald Murdoch. Cambridge: Cambridge University Press, 1984.

Dewey, John. *Reconstruction in Philosophy.* Rahway, NJ: Holt, 1920.

Goethe, J. W. *Goethe's Botanical Writings.* Translated by B. Mueller. Woodbridge, CT: Oxbow, 1989.

Hanby, Michael. *No God, No Science? Theology, Cosmology, Biology.* Chichester, UK: Wiley-Blackwell, 2013.

Heidegger, Martin. "The Question Concerning Technology." In *Basic Writings*, edited by David Krell, 307–42. London: Harper Perennial, 2008.

Jonas, Hans. *Philosophical Essays: From Ancient Creed to Technological Man.* Englewood, IN: Prentice Hall, 1974.

———. *The Phenomenon of Life: Toward a Philosophical Biology.* Evanston, IL: Northwestern University Press, 2001.

Lewis, C. S. *The Abolition of Man.* New York: Harper Collins, 2001.

18. For a fuller elaboration of this point, see Hanby, *No God, No Science?*, 9–45.

O'Connor, Flannery. *Mystery and Manners*. Edited by Sally and Robert Fitzgerald. New
    York: Farrar, Straus and Giroux, 1962.
Polanyi, Michael. *The Tacit Dimension*. Gloucester, MA: Peter Smith, 1983.
Schindler, David L. *Ordering Love: Liberal Societies and the Memory of God*. Grand
    Rapids: Eerdmans, 2011.

12

Wondering about the Science/
Scientism Distinction

D. C. Schindler

IT WOULD BE DIFFICULT to find a greater "champion of wonder" among contemporary philosophers than William Desmond. While it is common enough to signal the importance of wonder for philosophy—one can imagine that most introductory courses make at least passing reference to Plato's observation regarding the need for this experience in order to enter into the "big questions"—it is rare for one to think in depth about what this experience means, to reflect on the nature of wonder, on the significantly different types of wonder, what this origin implies about the nature of philosophy, the nature of knowledge, and the nature of being simply. And yet such reflection is required if we are to judge properly another claim one hears at least as often as the passage cited from Plato's *Theaetetus*, namely, that wonder is also essential to science.[1] From the depth of reflection that Desmond opens up—as we will see below—it becomes clear that wonder is not simply the same as a desire to know, and knowledge is not simply the same as the univocally determinate grasp of things that we tend to associate with science.

But there is another line of questioning that Desmond's essay prepares for but does not itself directly follow. I would like to suggest that this line of questioning is indispensable if the recovery of wonder that Desmond proposes is to bear all the fruit it promises. Early in the essay we are discussing

---

1. For a recent argument regarding the importance of wonder in human life, see Henderson, *A New Map of Wonders*.

here, Desmond observes that the critical questions he raises do not concern science per se, but "scientism," which he says is not science itself but rather an "interpretation"[2] of science—specifically, a "philosophical" one that takes science to offer a privileged view of the whole of reality. The distinction between science and scientism is a familiar one, which people typically invoke in order to preserve the evident value of the many things "science" has taught us and the many things it has produced to enhance and improve our life in this world, while enabling criticism of the presumptuous claims made on its behalf, or in other words of the tendency, in some scientists or even more commonly in popular and oversimplified translations of the work of science, to exaggerate science's field of competence. In this brief engagement with Desmond's essay, I would like to propose that the distinction between science and scientism is not as obvious as we might think, and that Desmond's own profound and nuanced account of wonder and knowing itself shows why this is so. More specifically, I would like to raise the question whether "scientism," far from being an aberration, is not rather a natural implication of what we normally mean by "science" in the contemporary world. To be sure, a full argument on this score, which would require both an extensive theoretical argument and a thorough study of historical sources, is not possible in this space. Instead of this, I intend here simply to unfold what I see to be implications of Desmond's excellent insights in relation to this distinction and the question I am raising about it.

Let us begin by drawing attention to one of the central points Desmond makes in this essay: it is not enough simply to affirm that the pursuit of knowledge begins with wonder; a more mindful attention reveals that there are different kinds of knowing, and these different kinds of knowing are correlated with different kinds of wonder. But it is also not enough simply to register the fact of a diversity of kinds in this matter. According to Desmond, we discover a privileged kind of wonder, which is wonder in its most original form, namely, astonishment. All other forms of human mindfulness have their roots in astonishment, and if there are indeed other, more determinate, kinds of wonder, these can be properly themselves only if they preserve in their own way the memory of their origin. Thus, the most radical kind of wondering is not a puzzlement about why things are the way they are, why they are one way rather than another; nor is it a mere curiosity about what things are, how they work, and what they mean. Rather, at the foundation of the whole is an astonishment, a being taken aback, by the existing of things simpliciter: an amazed wondering at the fact *that* things are, the being of things, which Desmond helpfully describes as

2. Desmond, *The Voiding of Being*, 97 [16].

the "too-much-ness" of things, their "overdeterminacy," their superabundance of meaning. This state of being, for Desmond, is radically receptive,
but without being simply passive:[3] the *passio essendi* [the passion of being]
precedes and makes possible the *conatus essendi* [the effort of being], the
striving after existence, and this original wondering is something *into which*
we are taken, rather than being a way of relating to things over which we
retain a certain sovereign control.[4]

   To be very brief here about matters that would require more subtle and
nuanced elaboration, the other basic kinds of wonder relate to this original astonishment, according to Desmond, as follows: *perplexity* begins as
alternative paths open up from this original experience, and one is beset
by uncertainty and doubt (the word "doubt" is, Desmond often points out,
etymologically related to the word "double"); a more active, and perhaps
even skeptical and critical, reasoning is engaged here insofar as judgment
is required, which involves deliberation and resolution. In the proper form,
as recollective of astonishment, perplexity allows our response to things
to grow in maturity; we come to ourselves in reasoning as we advance in
our appropriation of the matter at hand, and we keep from settling into the
complacency of an immediate, and superficial, apprehension. But severed
from the positive, life-giving root of astonishment, perplexity can become a
skeptical project, which systematically doubts in order to attain an absolute
certainty that no longer needs to receive from its other. Descartes is the obvious paradigm here. This kind of systematic doubt for the sake of mastery
is thinking as pure negativity; it affirms mediation only as originating in,
and terminating at, pure self-mediation (Descartes leads to Hegel).

   Finally, after astonishment and perplexity, there is curiosity, which is
a desire to move from indeterminate unknowing to a determinate grasp
of a thing. This is the passion that we associate most directly with science.
Desmond quotes Einstein in this context: "Never lose a holy curiosity."[5] The
point is that curiosity represents a desire to figure things out, which is to be
sure a kind of restlessness, but one that aims at definitive resolution as far as
possible. The tradition has recognized *"curiositas"*—understood as a trans-
gressive desire to ferret out all secrets and as an indiscriminate interest in
anything and everything, which weakens our particular and focused attentiveness to things that truly matter—as sinful.[6] If curiosity, so conceived, is a

---

3. Ferdinand Ulrich explains that, though wonder is something one undergoes, it
remains a "deed of the one who is struck." Ulrich, *Der Mensch als Anfang*, 146.

4. Heidegger also presents the experience of wonder in this way: see his *Basic
Questions of Philosophy*, §§37–38.

5. Desmond, *The Voiding of Being*, 116 [33].

6. The classic study of the notion of *curiositas* in classical literature is André

sin, it is because (like the perplexity that degenerates into sterile skepticism) it indicates a desire for domination, to be "master over." But the unbridled eagerness to know can also have a good and healthy form: the turning to things in curiosity, which opens us inquisitively to what they are, how they work, how they fit together, can deepen our astonishment, our holy respect for them. As Desmond puts it, "there can even be something of ontological and epistemic reverence in this turn to things. After all, it too participates derivatively in our original porosity to the astonishing givenness of being. Curiosity can release our sense of marvel at these given intricacies of things."[7]

Now, it is crucial to see that these three deeply intertwined forms of wonder are not merely subjective dispositions or attitudes. The assumption that they are is a very common misunderstanding. We think that preserving wonder, sustaining a genuine astonishment in the face of things, is simply a matter of keeping ourselves properly open, not letting ourselves become too complacent or habitualized, but instead constantly prodding ourselves to keep asking questions so that our desire to know does not die out. But Desmond's reflection helps us see that all of this strained effort, unless it is properly qualified, represents a kind of self-directed will that is the very opposite of wonder in its proper form; it becomes an expression of the *conatus essendi* rather than the more radical *passio essendi*. Desmond is clear that true wonder has an "ontological bite,"[8] as he puts it; it is not just a subjective disposition, but a sort of unsolicited breaking into the self by the reality of what is other. This is why Desmond insists that, in the question of wonder, the "senses of being are at stake."[9] The subjective and the objective dimensions of this experience are inseparable from each other: a certain kind of wonder arises, we might say, in response to the specific sense of being that reveals itself, and in relation to the different forms of knowing. We can be astonished only because being is in truth overfull, and only because it is the intrinsic nature of reality to reveal itself in an epiphanic way as "too much" for us to bear.[10] And if curiosity does not have to remain the desire existentially to dominate all of reality so that no shadow of a mystery would

---

Labhardt, "Curiositas: Notes sur l'histoire d'un mot et d'une notion." It is interesting to note that Pope Pius X identifies curiosity (and pride) as the essential cause of the heresy of "modernism": see *Pascendi Dominici Gregis*, 1907, 40.

7. Desmond, *The Voiding of Being*, 118 [35].

8. Desmond, *The Voiding of Being*, 107 [25].

9. Desmond, *The Voiding of Being*, 101 [20].

10. The end of Thornton Wilder's play *Our Town* presents this in an unforgettable way: "Emily: 'Does anyone ever realize life while they live it—every, every minute?' Stage Manager: 'No. Saints and poets maybe—they do some.'" Wilder, *Our Town*, 95–96.

elude the omnivigilant searchlight of our attention, it can only be because
the whatness of things and the distinct way they function and interrelate
are *inherently* surprising and meaningful. If the world, by contrast, were in
the end nothing but a collection of facts, of statements of what is the case,
as the early Wittgenstein seems to have proposed,[11] then astonishment, if it
existed at all, *would be* a mere "subjectivized" "feeling of 'gosh' or 'wow,'"[12]
and a self-initiated curiosity, aimed at mastery, would be the only serious
human disposition, as Descartes believed.[13] The question of wonder turns
on the question of being, the way reality *is*, and the way we come to have
access to it.

The reason this is important is that it reveals something often overlooked in discussions of wonder, namely, the importance of *the objective
dimension of things*, so to speak. The question of wonder, in other words,
concerns among other things ontological structures, our interpretation of
those structures, the intrinsic logic of our methods of knowing, the logic of
the instruments and technologies we employ to carry out those methods,
and so forth. To take the relevant example here, if we wish to open up the
resources of wonder in science it is not enough simply to remind ourselves,
as frequently as the work schedule allows, to step back from the microscope
and say "gosh" and "wow"—before getting back to work, which we do exactly in the same way as any other scientist, no matter how "scientistic"
he may be personally. Is "scientism" simply a personal matter, that is, an
attitude or private judgment made by an individual regarding the distant
implications of otherwise objective scientific "data," or does it concern the
very manner of investigating and the nature of the things that are investigated? Desmond's claim that "scientism is not science but an interpretation of
science"[14] prompts a further question in this respect: is it *simply* a post hoc
interpretation, contrived after the work of science is done, an interpretation
that presupposes that work as something given, complete in itself, which is
then just indifferently extrapolated in one direction or another? Is it the case
that we have the same basic "scientific facts," which are then either interpreted "scientistically" or instead interpreted properly scientifically, remaining the same basic facts regardless of their subsequent interpretation? What
we have seen so far suggests that this cannot be the case. The different kinds

11. The opening propositions of Wittgenstein's *Tractatus* are the following: "The
world is everything that is the case. The world is the totality of facts, not of things. The
world is determined by the facts, and by these being *all* the facts." See Wittgenstein,
Tractatus Logico-Philosophicus, 28.

12. Desmond, *The Voiding of Being*, 107 [25].

13. See Article 76 of René Descartes, *The Passions of the Soul*, 59–61.

14. Desmond, *The Voiding of Being*, 97 [16].

of wonder, and the different ways of knowing that correlate to them, have to preserve the memory of originary astonishment *within* themselves, both subjectively and objectively, in order to be healthy and genuinely human. In other words, the profound form of wonder has to be reflected in the very *form* of knowing in some analogous sense, and this means in the techniques and technologies, the instruments used in knowing, regardless of the kind of knowing that is at stake in a particular instance. And if astonishment is itself a response to the overdeteriminacy of being, the too-much-ness of things in the superabundance of their meaningful reality, then this implies that the mystery of being must in some sense bear on the very logic of every way of knowing, including those operating within the discipline of science.

It is not difficult to distil from this claim a particular question regarding the nature of science, as it is generally understood and practiced in the world today, even if the task of providing an even relatively adequate answer to the question is quite daunting, and beyond our scope. A positive formulation of the general line of inquiry would be something like the following: What would a science that understood itself as ontologically and methodologically guided by the mystery of being as overfullness *look* like? Is such a science anywhere in evidence today? Is it evident in any of the recognized founders of so-called modern science in its revolutionary break from the old "philosophy of nature" that understood itself as a sub-branch of theoretical science, ultimately governed by metaphysics, the science of being as being? How do figures such as Bernardino Telesio, Francis Bacon, Galileo, Boyle, Newton, and so forth, appear when measured by the standard of "astonishment"? (This is not meant to be a rhetorical question, but a real one: it would be worthwhile working through the writings of these figures with the question of "astonishment," as Desmond presents it, in mind.) Beyond the founders, are there practitioners today of science who integrate astonishment in their understanding of the nature and standard practice of science, or indeed in their understanding of the nature of things, precisely as that nature is revealed in the practices of the science in question? If there are such practitioners, what is their relationship to the mainstream, and how have they generally been received? These are all of course massive questions, which do not admit of hasty answers, but it should be clear that the issue Desmond has raised in his essay requires that they be pursued in a serious way.

For his part, Desmond says very little here about what genuine science is, and what it would look like in contrast to the "scientism" that he distinguishes from it, and he does not take up of course any of the more historically based questions I formulated, since these sorts of questions unfold in an evidently different register. But it is worthwhile attempting to sketch out at least some parameters of an answer on the basis of things he does say.

As mentioned above, Desmond states here that "scientism is not science but an interpretation of science." Specifically, scientism interprets science as "a philosophical interpretation of the whole, though it takes place within the whole."[15] The interpretation of the whole from within the whole is not a problem in principle, because in fact, Desmond says, it is impossible to explore things scientifically without, either explicitly or (more often) implicitly, presuming some interpretation of the whole. What is distinctive, and problematic, about scientism is twofold: first, it presumes science, on its own, to have "the resources to supply all answers to all the meaningful questions," and indeed, "the privilege and the right to make this claim."[16] Second, as our previous observations would suggest, this presumption or subjective disposition has an objective correlate, that is, a decision concerning the nature of things in general: "in various forms of scientism the underlying presupposition is that being is determinate, and in a manner that invites science to make it as univoccally precise as possible."[17]

Now, one might say that genuine *science*, in contrast to the scientism so characterized, would insist that it can answer only a limited set of questions, and that there are larger questions—spiritual questions concerning the meaning of life, the nature of love, and so on—that lie outside of its competence; moreover, if this is the case, it is because only *part* of being is determinate in the univocal sense just mentioned, while the rest of being, the other parts of being, so to speak, retain the dimension of mystery. But, given the reflection so far, this formulation cannot help but make us uneasy. Things cannot be so easily partitioned out as this formulation would suggest. The three kinds of wonder are not separate attitudes, applied to different parts of being; instead, astonishment is originary, and as such is meant so to speak to run through the whole and accompany the other layers of engagement. In this case, *every* exploration of being, no matter how particular and self-professedly partial, would have to embody astonishment in itself, in its own distinct mode of investigating things, in its methods and instruments, and in its conception of its proper object and its purpose.

There are a number of things it seems we could say, however schematically in this context, on the theme of science as an expression of astonishment: there is a difference between analyzing a whole into its parts for the sake of getting a better grasp of them, on the one hand, and absolutizing the

15. Desmond, *The Voiding of Being*, 97 [16].

16. Desmond, *The Voiding of Being*, 97 [16].

17. Desmond, *The Voiding of Being*, 101 [20].

parts in their isolation, on the other. What does "absolutizing" mean here? It means bracketing out what lies beyond the parts as irrelevant to their meaning, and assuming the parts are intelligible simply in themselves without reference to their larger context. Properly speaking, the light of intelligibility, just like the living source of wonder, comes from above and from below at the same time: we learn something about a thing by "dissecting," through analysis, but we do not learn everything; the very intelligibility that comes to light seeks resolution in the whole of which it is a apart. It is this seeking, this intrinsic movement of parts toward an integrating whole, I suggest, that opens scientific curiosity to true astonishment. A genuine science, which thus embodies real astonishment and not just a raw and relentless curiosity, will tend to think *holistically*. "Holistic" thinking, contrary to common assumptions, is not incompatible with detailed analysis and rigorous methods; it simply insists on viewing the parts, not as separate "bits," but precisely *as* real parts, and therefore *always*, at *every* level of analysis, expressions of a reality greater than themselves. This means, too, that science will naturally look beyond itself, precisely *as* science, and not simply when the hard work of science is over or before it begins, to the larger visions of philosophy, art, and religion, for insight into its *own* endeavor. Such an approach would be a science that avoids the totalitarian tendencies of scientism.

Whatever such a science might look like, we can be fairly confident it would be significantly different from what we generally take to be science today. This is not because there is a lack of individuals with an inclination to wonder seriously, so to speak, in the manner we have been discussing. To the contrary, if "man by nature desires to know," it is because man is most fundamentally the wondering animal, the animal given to astonishment. We can thus assume that "initially and for the most part," as Heidegger would say, people are inclined to receive the gift of being in this profoundly human way. But we have seen that the issue is not the subjective disposition alone; the logic of the activity itself and the nature of the reality in which it occurs is deeply relevant. When the activity at issue, moreover, is one that so fully dominates the public square, as it were, one that exercises such a clear authority, not only in the official arena of politics, but in the popular imagination more generally, then the question of how properly to investigate the nature of things does not concern scientists alone, but all of us. Desmond's profound essay, opening the resources of a mindfulness that remains true to the gift of being, ought to be received as an appeal addressed to us all, because it concerns what makes us genuinely human.

Bibliography

Descartes, René. *The Passions of the Soul.* Indianapolis: Hackett, 1989.

Heidegger, Martin. *Basic Questions of Philosophy: Selected "Problems" of "Logic."* Bloomington, IN: Indiana University Press, 1994.

Henderson, Casper. *A New Map of Wonders: A Journey in Search of Modern Marvels.* London: Granta, 2017.

Labhardt, André. "Curiositas: Notes sur l'histoire d'un mot et d'une notion." *Museum Helveticum* 17.4 (1960) 206–24.

Pius X. *Pascendi Dominici Gregis.* 1907. Online: https://www.vatican.va/content/pius-x/en/encyclicals/documents/hf_p-x_enc_19070908_pascendi-dominici-gregis.html.

Ulrich, Ferdinand. *Der Mensch als Anfang: Zur philosophischen Anthropologie der Kindheit.* Einsiedeln: Johannes Verlag, 1970.

Wilder, Thornton. *Our Town.* New York: Samuel French Acting Edition, 2013.

Wittgenstein, Ludwig. *Tractatus Logico-Philosophicus.* Mineola, NY: Dover, 1998.

13

Basil and Desmond on Wonder and the Astonishing Return of Christian Metaphysics

Isidoros C. Katsos

Liturgical and Doxological

On the eastern part of the Greek island of Crete, in the semi-arid area of Lasithi, lies the old monastery of Toplou. The monastery hosts the unique icon *Megas ei Kyrie* ("Great are you, O Lord"), written on wood by Ioannis Kornaros in 1770.[1] The name of the icon is not that of a person but of a doxological praise; this in itself is indicative of a very unusual theme. The central motif is the baptism of Christ, which is celebrated in the East on the feast of Theophany (January 6). Upon closer inspection, the central motif is placed *between* heaven (the Trinity and the angels) and earth (Mary with the Christ Child, Adam and Eve), whereas the descent of Christ into the underworld is depicted at the bottom of the icon. Fifty-seven episodes from the Old and the New Testaments surround the central motif on the right and on the left, all connected by running waters. The total of sixty-one motifs are enumerated by the artist and glossed by headings, which faithfully reproduce the verses of the *Theophany Poem*, attributed to the Patriarch Sophrony of Jerusalem. The poem contains the prayer *Megas ei Kyrie* ("Great are you, O Lord"), which is the central liturgical event of the

---

1. For a detailed description of the icon and its history, see Kiriaki-Sfakaki, *Megas ei Kyrie.*

169

**The Toplou icon, *Megas ei Kyrie* ("Great are You, O Lord"), 1770.
Painted on wood by Ioannis Kornaros.[2]**

annual Theophany feast. The first verse, engraved at the feet of the praying patriarch, reads as follows:

> Great are You, O Lord, and astonishing (*thaumasta*) are
> Your works, and no word will suffice to praise Your wonders
> (*thaumasiōn*).

2. Image source: Wikimedia Commons. https://commons.wikimedia.org/wiki/File:
Megas_Ei_Kyrie_from_Ioannis_Kornaros_1770.jpg. Accessed May 25, 2022.

The secondary themes of the icon narrate these wonders: the four elements, the stars, the lights and the four seasons, plants, fish, and animals,
all the way up to the human being. The narrative sequence culminates in a
fourfold vision of Christ: his birth, his baptism, his descent into Hades, and
his glorification in heaven. The narrative of the icon is therefore Christocen-
tric, while the main theme is baptismal, liturgical, and doxological. Thus, in
the words of the *Theophany Poem*: "All creation sang your praises upon your
appearing."

The theme of God's greatness contemplated in the wonders of creation is thoroughly biblical. It is how the poetic and prophetic traditions
unpacked the meaning of Moses' creation narrative: in astonishment and
wonder,[3] we contemplate God's glory through his works.[4] Early Christianity
emulated the same teaching:

> Great and astonishing (*thaumasta*) are Your works, Lord God
> Almighty! . . . Who shall not fear You, O Lord, and glorify Your
> name? (Rev 15:3, 4)

According to the scriptural teaching, the physical world is full of
wonder. Being struck by the wonder of the work, we glorify its Creator in
astonishment. Scripture teaches us to experience the world like a work of
art. As biological creatures, we live in the world, as parts of a great work of
art. As rational creatures, however, we have the capacity to look at the world
as if from the outside, as if stepping back from a great work of art. It is this
experience of being in-between, as both within and beyond the world, that
the icon of the Toplou monastery aims to incite in the spectator. At that
very moment, when we are suddenly transferred into the noetic space of the
in-between, acquiring a double vision of the world as if from the inside and
yet beyond, we cannot help but exclaim in doxological astonishment: *Megas
ei Kyrie* ("Great are you, O Lord").

Doxological and Metaxological

In "The Dearth of Astonishment," William Desmond offers a critical reflection on the nature of wonder.[5] The liturgical and doxological experience
of the world lies at the heart of his analysis: "Great artworks, like religious
reverence or awe, may offer us striking occasions of originating wonder

3. Pss 40:5; 139:14.
4. Ps 19:1; Exod 15:3.
5. Desmond, *The Voiding of Being*, 96–125 [15–42].

. . . ."⁶ Desmond distinguishes between three modalities of wonder: astonishment, perplexity, and curiosity. Astonishment, the first modality, is "a
kind of amazement," "a kind of ontological stupor." We are stricken by it
every time we become aware of the original porosity, the otherness, openness, and givenness of being. Desmond writes: "The otherness seems to stun
us, bewilder, even stupefy us."⁷ Perplexity is the second modality of wonder.
It is the bewildering, the doubt and the uncertainty that beset us when we
are awakened to the perturbing ambiguities of transient being and to the
experience of the world as an enigmatic and intimating matrix of *aporias*.
Desmond continues: "Something puzzles us and we cannot quite solve the
puzzle."⁸ Through introspection, perplexity accompanies the awareness of
our transient nature, "of ourselves as the most baffling of beings."⁹ Curiosity
is wonder's third modality. It is the inquisitiveness with which we approach
the world; it is our desire to know things through determinations.

Desmond constructs the difference between these three modalities as
a difference in intentionality. Astonishment is primitive and pre-intentional.
It is our openness towards the maternal porosity of being before the mind
begins to shape any intention or desire to know. It is "an original porosity of being in us, an original porosity that is open to being."¹⁰ Curiosity,
our intentional desire to know, insofar as it is already a specification of that
original porosity, takes place in astonishment and derives from it. Perplexity, on the other hand, dialectically oscillates between the unintentional
and the intentional. Becoming aware of its individuated being, the mind
is bedazzled by its transient nature. Such confusion is a first remove from
astonishment, whereas curiosity is a second remove. As the mind desires to
know, its motion becomes intentional, and it yearns to take shape. Science
is the archetypical realm of curiosity. In moderation, scientific curiosity allows us to know things in their details. In excess, however, curiosity turns
science into "scientism," that unquenched needs "to supply all answers to all
the meaningful questions."¹¹ Once driven by excessive curiosity, the mind
becomes aggressive and enquiry mutates into inquisition, an exercise
in ontological violence. In its endless pursuit of the master key that will
unlock all secrets, the dream of "a theory of everything," the mind loses
sight of the simplicity of being, of being as sheer presence, and of the sense

6. Desmond, *The Voiding of Being*, 119–20 [36].
7. Desmond, *The Voiding of Being*, 107 [25–26].
8. Desmond, *The Voiding of Being*, 112 [29].
9. Desmond, *The Voiding of Being*, 113 [30].
10. Desmond, *The Voiding of Being*, 99 [18].
11. Desmond, *The Voiding of Being*, 97 [16].

of astonishment that it incites. In displacing astonishment, what is left of curiosity is "a contraction of wonder . . . ."[12] It is this "dearth of astonishment" that Desmond bemoans and regrets in any totalizing, "scientistic" worldview.

The complaint is anthropologically and existentially motivated. As Desmond explains, "How we understand the culture of knowing is in question."[13] Here, culture is meant in the deepest and widest sense. What is at stake is what it means to know. In vindicating astonishment over against perplexity and curiosity, Desmond wants to safeguard the possibility of mindfulness, our childlike—not childish!—relation to the world.[14] He advocates for a mode of wonder in which we may open up ourselves to the fullness of being. In this mode of wonder, we may experience the joy of sheer presence before the mind is driven away by an endless sea of determinations. Determinations provide answers and dispel wonder; they seem to provide security. But this is an illusion. Being manifests itself through endless determinations. The mind that desires to discover them all sets for itself an impossible task. Its curiosity becomes the source of its "infinite restlessness."[15]

This seems to be Desmond's deepest insight. Knowing is being. Our thirst for cognition through determinations is chimeric because it aspires to reduce the irreducible and to determine the overdeterminate. So long as curiosity "spends itself in ceaseless accumulation of determinate cognition(s),"[16] our desire to know becomes the deepest source of our existential agony. What is at stake, then, is the equanimity of our being-in-the-world. Astonishment, as a cognitive mode of pre-intentional openness to the porosity of being, is a "step back" from determinate to overdeterminate being. We are determinate beings, however. In our experience of the overdeterminate, being astonished, we become displaced. We are no longer "there" or "here," but there-and-here, there-and-beyond. It is like standing in front of the Toplou icon. In the act of contemplation, the spectator of the icon and of the world awakens in the noetic space of the in-between (*metaxu*). Suddenly, in the astonishment of wonder, the world becomes the icon and we become metaxological: we are inside and yet outside the painting, we are there-and-here, there-and-beyond. Doxological is metaxological:

12. Desmond, *The Voiding of Being*, 120 [37].
13. Desmond, *The Voiding of Being*, 99 [18].
14. Desmond, *The Voiding of Being*, 111 [28].
15. Desmond, *The Voiding of Being*, 121 [38].
16. Desmond, *The Voiding of Being*, 101 [25].

being-there-and-yet-beyond, we find ourselves in the world and yet not of the world.[17]

Metaxological and Metaphysical

Let there be no doubt. Desmond's aim is no less ambitious than to retrieve the ancient project of metaphysics for postmodernity. His *metaxu* denotes being "between indeterminacy and overdeterminacy."[18] His concern is that "the senses of being are at stake."[19] If this is the case, then we stand once again in front of the traditional concerns of metaphysics. The language of "determinacy" is not difficult to decipher. It leads us directly back to Hegel, for whom "determinateness" denotes "quality."[20] Determinate being is qualitative being. Pure being is being without any further quality or determination. In its simplicity, pure being is infinite. It is the absolute being—Desmond's overdeterminate. In its self-negation, pure being is nothingness. Deprived of all quality, it is Desmond's indeterminate. Determinate being occupies the middle position between the overdeterminate and the indeterminate. Transient being, "becoming," is always determinate being, the union of the overdeterminate and the indeterminate. We are born as determinate, transient beings, yet we are destined to become metaxological. We are called to displace ourselves from our self-determinacy and become porous to the excessive "too-muchness" of being that surrounds us.[21] Ontological astonishment calls us to affirm our role as *metaxu*, as in-between two aspects of infinity, the overdeterminate and the indeterminate, and to choose being in its hyperbole rather than in its formlessness.

Desmond's metaphysical language is, however, much older than that of Hegel. The basic categories of the determinate, the indeterminate, the self-determining, and the overdeterminate, are in fact a distant echo of the fourfold division of being in the *Philebus*: the *peras* (limit), the *apeiron* (unlimited), their mixture (*meiktēn kai gegenēmenēn ousian*), and its cause (*meixeōs aitian kai geneseōs*).[22] As for "metaxology," it is another way of speaking of participatory ontology. We stand in front of ancient categories of thought recast in slightly new, yet clearly suggestive terms. That is not to say that the project of metaphysics from Plato onwards is one and the

---

17. Cf. John 17:14–18.
18. Desmond, *The Voiding of Being*, 102 [21].
19. Desmond, *The Voiding of Being*, 101 [20].
20. See here and in what follows, Hegel, *The Science of Logic*, 45–124.
21. Desmond, *The Voiding of Being*, 108 [26].
22. Plato, *Phil.* 23c-27b.

same. That would be a gross misunderstanding. It is to say, however, that the project of metaphysics throughout its long history has always been *about* the same thing, namely, the question of being. If so, the senses of "being" are always at stake every time we enquire into the meaning of being, for, as Aristotle remarked, "being is said in several ways."[23] The project of metaphysics is about being as it is said in the most primary way.

There are multiple ways in which one may understand the project of metaphysics. One may understand it as the task of disambiguation of the various senses of "being" and therefore of identifying metaphysics with ordinary-language philosophy. Desmond's critique is that this approach cannot be critical enough: it presupposes intentionality, determinations, subject-predicate relations, whereas our primary experience of being precedes intentionality. Metaphysics as a linguistic project therefore misses the most important—and salvific—cognitive mode of being: wonder as astonishment. Another way of understanding the project of metaphysics is by distinguishing primary being from non-primary being and by identifying the primary way in which being is said with primary being. Metaphysics would then be the science of primary being, while the derivative ways in which being is said would be the subject matter of the particular sciences. On this approach, the project of metaphysics becomes the project of the discovery of a "master science," the science of being as a whole, of "being *qua* being." Desmond also reacts against this approach insofar as it leads to a double impasse: first, it risks stripping off primary being from any intelligibility until it has become determinately intelligible; and second, it risks culminating in a totalizing view of being as self-determination. But cognition of the whole through (speculative) determination is the chimeric dream of scientism.[24] The heart of the issue with metaphysics as the science of all sciences is that it misses the most primary notion of being, the one that calls forth our ontological astonishment. It is against this "dearth of astonishment" in speculative metaphysics that Desmond emphatically reacts.

Metaphysical and Theological

There is another way of understanding the project of metaphysics: metaphysics as a project animated by wonder. Aristotle is quite explicit about the foundational role of wonder in the question of being, although his words are often conveniently forgotten. It is helpful to remind ourselves of them

23. Aristotle, *Met.* IV.2 1003a33–34.
24. Desmond, *The Voiding of Being*, 98, 103–6 [17, 21–24].

in full. When stating his overall aim in the central books of the *Metaphysics*, Aristotle remarks:

> Indeed, that which is always, both now and long ago, sought after and which is always a source of puzzlement [*to aporoumenon*], i.e. the question, *What is being?*, is really the question, *What is primary being? [ousia]*. . . . So we too must, most of all, primarily, and so to speak exclusively, investigate about that which is being in this way [i.e. that which is being in the primary way]: what is it? (*Met.* VII.1 1028b2–7)[25]

The word that Aristotle uses for wonder is *aporia*, which means "puzzle." He seems to have two senses of *aporia* in mind. According to the reading that I follow here, Aristotle speaks of *aporia* in a primary and in a derivative sense. The primary sense is "the mental state of puzzlement and perplexity." The derivative sense is that of "particular puzzles and problems which . . . are responsible for and the cause of the mental state of *aporia*."[26] On this reading,

> Aristotle thinks that the question, "What is being?," is not just any kind of question, but an *aporia*; i.e. a problem and puzzle that presents itself to us and makes us puzzled about it. This fascinating view arguably goes back to Plato, especially in the dialogue *Sophist* (242b-243d), where Plato argues that we are thoroughly perplexed and puzzled about being.[27]

Indeed, a careful re-reading of the *Metaphysics* reveals that the question of being, or, more precisely, the question "What is it for something to be?," is pursued in a Socratic manner. Its proper method is *aporia*-based. Our thinking about being in general is prompted by the puzzles that strike and baffle us about it.[28] Let us remember Aristotle: "that which is always sought after and which is always a source of puzzlement, i.e. the question, 'What is being?'" The question of being is *always* a source of wonder and of puzzlement.

That is not all. While we remain within the realm of *aporia*, we are still within the realm of determination and Desmond's critique about the dearth of astonishment still holds. Let us remember how Aristotle continues: ". . . the question, 'What is being?', is really the question, 'What is primary being?'"

25. All translations from Aristotle's *Metaphysics* are from Politis, *Aristotle and the Metaphysics*, whose *aporia*-based reading I follow.

26. Politis, *Aristotle and the Metaphysics*, 69.

27. Politis, *Aristotle and the Metaphysics*, 6.

28. Aristotle, *Met.* III.1 995a24–25; 995a34-b2.

Aristotle's project at first appears "philosophical" because it follows the Socratic method. Gradually, it reveals itself to be "theological." The search for being as a whole and *qua* being leads Aristotle to the most fascinating discovery of all—God. It is here, in book XII of the *Metaphysics*, which should not be disconnected from the project of metaphysics on the pain of destroying its coherence,[29] that puzzlement and perplexity are transformed into astonishment. This God of *Metaphysics* is deeply paradoxical, being completely changeless and yet internally active. This paradoxical God no longer generates perplexity and confusion. Instead, God incites a breath-taking sense of awe and wonder being "good," "perfect," "beautiful," and "beloved."[30] This God moves everything by desire, by being himself the ultimate object of love.[31] Being utterly loved, the God of *Metaphysics* no longer causes puzzlement but astonishment. God's perfect and primary mode of being is astonishing (*thaumaston*) and beyond astonishment (*thaumasiōteron*).[32] Astonishment is the proper mode of cognition of primary being. It may well be the case that the project of metaphysics proceeds dialectically through puzzlement (*aporia*; *aporein*). However, it is also the case that metaphysics as a project, the quest for the meaning of being, begins and ends in astonishment (*thaumaston*; *thaumazein*). Philosophy arises out of astonishment and ends in astonishment—astonishment as the primary modality of wonder.[33] It is this approach to metaphysics as astonishment that Desmond valiantly aims to retrieve for our times.

Theological and Doxological

This approach to metaphysics is not to be found in traditional textbooks, although, astonishingly, it can be found in the tradition of the early church. From Philo onwards, in the so-called "Alexandrian tradition," the God of Genesis and of Exodus is identified with the God of *Metaphysics*: the Creator God,[34] He-Who-Is,[35] is primary being,[36] true being,[37] unqualified being.[38]

29. For the argument, see Politis, *Aristotle and the Metaphysics*, 295–97.

30. Aristotle, *Met.* XII.7 1072a26–28.

31. Aristotle, *Met.* XII.7 1072b3–4.

32. Aristotle, *Met.* XII.7 1072b25–26.

33. Plato *Theat.* 155d. Aristotle *Met.* I.2 982b 12–19; 983a12–21.

34. Philo, *Opif.* 171 (et passim): *ho dēmiourgos*, with reference to Gen 1:1.

35. LXX Exod 3:14: *ego eimi ho ōn*.

36. Philo, *Opif.* 172: *ho ōn ontōs*.

37. Philo, *Abr.* 143: *ho pros alētheian on*.

38. Philo, *Mut.* 9: *ison tōi einai*.

Conversely, primary being is the perfectly good and beautiful God,[39] the God worthy of love[40] and acting out of love.[41] The identification of the God of the *Pentateuch* with the God of *Metaphysics* has often been attacked and misunderstood because the most basic tenet of the project of metaphysics—that all philosophy begins and ends in astonishment—has been ignored, forgotten and suppressed by alternative readings that reduce wonder to its second and third modalities, perplexity and curiosity. A correction is badly needed. It is misleading to speak of "identification" since identification assumes that there is a "God of Exodus" and a "God of *Metaphysics*" and that the two are isomorphic. There is much nonsense in such an assumption. Scripture does not allow for more than one god, a "God of Scripture" and a "God of philosophy." Nor is the God of *Metaphysics* an isomorphic being. Instead, this God explains and exemplifies being in the most primary way, transcendentally, i.e. separately from all sensible determinations.[42] There are no two gods, a "God of Scripture" and a "God of philosophy," to compare and identify. There is only one and the same paradoxical God of love who is the answer to the prayer of the faithful *and* to the search of the philosopher. God is what God is. God is what primary being is: the supreme object of love and at the same time the ultimate explanatory principle of all that is. The primary modality of wonder is the proper mode of cognition of God *qua* God *and* of being *qua* being because God *is* primary being. Astonishment *is* wisdom. Piety *is* true philosophy. Theology *is* pious ontology. The cornerstone of piety *is* the cornerstone of wisdom. It is God who calls forth our ontological astonishment. Plato, Aristotle, and their epigones seem to have known this. The church fathers also seem to have known this. It is good for contemporary Christians to remember where they come from. Desmond helps us remember that there is an alternative approach to the project of metaphysics, an approach that safeguards the doxological aspect of being because it refuses to reduce the primary sense of wonder to perplexity and curiosity. The return of astonishment in metaphysics heralds the astonishing return of Christian metaphysics.

We see the metaphysics of astonishment at play throughout patristic literature. Basil of Caesarea offers an eloquent example. His celebrated *Hexaemeral Homilies*, a series of nine sermons on the biblical creation narrative, contain a sustained disquisition on the metaphysics of wonder. In the homilies, we find no less than thirty-two references to the language of

39. Philo, *Opif.* 136: *ho gar alētheiai kalos kai agathos.*

40. Philo, *Cher.* 7: *philotheos.*

41. Philo, *Deter.* 13: *anthrōpōn theophilōn.*

42. Aristotle, *Met.* XII.7 1073a 4–5: *ousia . . . kekhōrismenē tōn aisthētōn.*

"astonishment" (*thauma*) and to the mental state that astonishment incites (*thaumazō*) in the pious spectator of creation.[43] Contemplation of nature begins in wonder and ends in wonder. Like all genuine philosophy, contemplation of nature arises out of wonder. Natural theology is Christian philosophy:

> *In the beginning God created the heavens and the earth.* Astonishment (*to thauma*) at the thought checks my utterance. What shall I say first? Whence shall I begin my narration? (*Hom.* I.2)[44]

Like all genuine philosophy too, contemplation of nature ends in a state of wonder. True philosophy is doxology:

> What words can express these marvels? What ear can understand them? What time can suffice to say and to explain all the wonders of the Creator (*tou tekhnitou ta thaumata*)? Let us also say with the prophet: *How great are thy works, O Lord? Thou hast made all things in wisdom* [Ps 103.24]. (*Hom.* IX.3)

Like the Socratic project of metaphysics, our enquiry into being is *aporia*-based.[45] Yet, as we are stricken by the puzzles of being, we enquire in a state of puzzlement and confusion. In the cognitive mode of perplexity, the mind cannot find rest:

> Therefore, I urge you to abandon these questions and not to inquire upon what foundation it [namely, the earth] stands. If you do that, the mind will become dizzy (*illiggiasei hē dianoia*), with the reasoning going on to no definite end. (*Hom.* I.8)

Dizziness is the sign of excessive curiosity and the insatiable desire for determinate cognition that comes with it:

> These same thoughts, let us also recommend to ourselves concerning the earth, not to be curious (*polypragmonein*) about what its substance is; nor to wear ourselves out by reasoning (*katatribesthai*), seeking its very foundation; nor to search for some nature destitute of qualities, existing without quality of itself . . . . (*Hom.* I.8)

---

43. Based on a *Thesaurus Linguae Graecae* search (last accessed: May 25, 2022).

44. All translations of Basil's *Homilies* are from Basil of Caesarea (1963), except when indicated.

45. Basil, *Hom.* I.9.21: *aporia*.

This is a deceptive path of enquiry into being. It leads not to equanimity but to perpetual agony by reducing wonder to curiosity and being to determinations. True being, however, is overdeterminate, hence beyond comprehension:

> Set a limit, then, to your thoughts, lest the words of Job should ever censure your curiosity (*polypragmosynē*) as you scrutinize things incomprehensible (*ta akatalēpta*) . . . . (*Hom.* I.9)

If we want to understand the world that we inhabit, we should not seek knowledge through determinations but wonder in astonishment: "For it is proper to wonder (*thaumasai*) that . . . ."[46] This is the proper cognitive mode of engagement with the world. Just like Desmond, Basil reacts emphatically against all forms of scientism that result in the dearth of ontological astonishment. On the one hand, he warns us against the speculative project of metaphysics as the science of all sciences.[47] This is a project driven by insatiable curiosity.[48] On the other hand, he also warns us against the dialectical scientism of alternative totalizing worldviews.[49] They result in endless theories.[50] By contrast, the proper mode of enquiry is one that calls forth our ontological amazement (*ekplēxis*)[51] about the world we inhabit. It is the astonishment (*thauma*)[52] caused by the overdeterminacy of being. In anticipation of Desmond—or rather in sharing the same approach to the project of metaphysics—Basil too invites us to become *metaxu*, to be stricken by the astonishment of an ever-present "too muchness," and to displace our awareness from the formlessness of the indeterminate to the beauty of the overdeterminate infinite:

> Let us glorify the Master Craftsman for all that has been done wisely and skilfully; and from the beauty of the visible things let us form an idea of Him who is more than beautiful; and from the greatness of these perceptible and circumscribed bodies let us conceive of Him who is infinite (*apeiron*) and immense and who surpasses all understanding in the plenitude of His power. For, even if we are ignorant of things made, yet, at least, that which in general comes under our observation is so astonishing

46. Basil, *Hom.* I.7.
47. See Basil, *Against Eunomius* I.12–14 [English translation, 108–14].
48. See Basil, *Against Eunomius* I.14: *polypragmosynēs hōs anefiktou*.
49. Basil, *Hom.* I.11: *pithanologian . . . dielysen . . . oikeian anteisēgagen doxan*.
50. Basil, *Hom.* I.11: *adoleskhian*.
51. Basil, *Hom.* I. 10.
52. Basil, *Hom.* I. 10.

(*thauma*) that even the most acute mind is shown to be at a loss as regards the least of the things in the world, either in the ability to explain it worthily or to render due praise to the Creator, to whom be all glory, honor, and power forever. (*Hom.* I.11, translation slightly amended)

It is this ever-recurring and ever-elusive project of metaxological metaphysics, the metaphysics that begins and ends in astonishment, in *thaumazein*, that transforms the world into an icon, awakening us into the noetic space of the in-between, outside and yet inside the icon-world. It is in this mode of wonder that we finally realize that the metaphysical *is* theological, an experience only expressed through the doxological coda of liturgical language:

> Great are You, O Lord, and astonishing are Your works, and no word will suffice to praise Your wonders.

Bibliography

Basil of Caesarea. *Against Eunomius.* Translated by Mark Del Cogliano and Andrew Radde-Gallwitz. Washington, DC: The Catholic University of America Press, 2011. [*The Fathers of the Church: A New Translation* 122].

———. *Exegetic Homilies.* Translated by Sr Agnes Clare Way, C.D.P. Washington, DC: The Catholic University of America Press, 1963. [*The Fathers of the Church: A New Translation* 46].

Desmond, William. *The Voiding of Being: The Doing and Undoing of Metaphysics in Modernity.* Washington, DC: The Catholic University of America Press, 2020.

Hegel, Georg Wilhelm Friedrich. *The Science of Logic.* Edited and translated by George di Giovanni. Cambridge: Cambridge University Press, 2010.

Kiriaki-Sfakaki, Athina. "Megas ei Kyrie" 2013: Kiria Akrotiriani, 1770, Ioannis Kornaros (Irakleio: Itanos) [in Greek: Αθηνά Ν. Κυριακάκη–Σφακάκη, «Μέγας εἶ Κύριε»: Κυρία Ακρωτηριανή ΑΙΨΟ (1770), Ιωάννης Κορνάρος (Ηράκλειο: Ἴταμος, 2013)].

Politis, Vasilis. *Aristotle and the Metaphysics.* London: Routledge, 2004.

PART THREE

14

The Children of Wonder

ON SCIENTISM AND ITS CHANGELINGS

William Desmond

MY SINCEREST THANKS FOR the diverse responses of the different contributors. I am honored to be taken with such seriousness and given such illuminating responses. In his thoughtful introduction Paul Tyson offers a helpful overview and summary of the different contributions and I will turn to these contributions shortly. These contributions are very rich and occasion much thought. I am grateful for the care taken, for the attentive generosity of interpretation, for many illuminations in receiving the insights of my interlocutors, for intelligent fidelity to "the between" as it offers itself for companioning thinking. I will try to offer a relevant thought or two in response, while being aware that much more might have been said, so rich are they each as invitations for further reflection. Some of my remarks are inspired by the originals, and if supplementing, in all cases I owe a debt of gratitude. In response to a plurality of voices ones own responding voice courts a plurivocity. The companioning voices are sometimes more systematic, sometimes more poetic, sometimes between these two and letting them be porous to each other. Each is a singular contribution, yet in the diversity there is convergence on common themes. While trying to give each contribution its due, I will here and there indicate what struck me as companioning communication between them.

Being diverse, these contributions call forth different responses. Some are thoughtful and illuminating illustrations of some of the notions I offer. Some are amplifications of these notions in also illuminating directions. Some are worried about the sustainability of the distinction of science and scientism. Generally, the responses show hearteningly the fertility of these notions, while others remind one of the matrix of the equivocal within which science itself is birthed and grows. That matrix tempts me to think of the parable of the wheat and the darnel that are twined together and grow together but we have to approach this twining with finesse lest we destroy the nourishing growths with the stifling. I will say something about the darnel in the field of science and perhaps indeed the overgrowth of scientistic darnel to the detriment of life-supporting wheat.

I have entitled my remarks as I have done because the figure of the child recurs in a number of contributions and, noting this, I see a connection with scientistic curiosity and the changelings of wonder. A changeling in lore is a replacement for a stolen child and resembles almost exactly the child stolen, except that the living singularity is dulled or extinguished. The replacement seems exactly the same and is exactly not the same: a case of univocity that is entirely equivocal. Are scientistic curiosities the changelings of the living children of wonder?

I ask because I see around us a significant proliferation of counterfeit doubles and I see this as bound up with a dearth of astonishment. While Nietzsche spoke of nihilism as the self-reversal of the highest values, I would say our nihilism is rather the production of counterfeit doubles of the highest values. Such counterfeit doubles can colonize all the domains of life, including the seeking for true knowledge. I think scientism gives expression to a kind of counterfeit double of science. Remember counterfeits claim and gain their credit worthiness by seeming to *perfect* the original. The counterfeit double mimics an original and does so in a consummate degree such that we often cannot see any difference between the original and the counterfeit. The counterfeit seems to perfect the original. Essential differences vanish in a seeming movement towards a better or more perfect version.

An example I like to cite as illustrative is how during World War II the Germans produced counterfeits of British £20 notes that were so good they were hard to detect, and for a time they caused mischief. The counterfeits finally were identified by the fact that they were *more perfect* than the true notes. The counterfeits had no flaws, whereas the original currency notes always had some small flaw. While the counterfeit presented itself as more perfect, the original even in its imperfection was the truly creditworthy currency. The counterfeit was backed by nothing trustworthy and true except perhaps superior techniques of currency production. In truth, the imperfect was more perfect.

Let me cite one or two instances with some relevance to scientism. Consider claims that artificial intelligence will produce the perfection of intelligence. Would this not be mindless mind, sourced in geometry without finesse, claiming to perfect intelligence? Is not *such perfected intelligence stupid*, lacking finesse for the first and last things? One might have a more and more perfected calculation, but in a true sense it would always be mindless, devoid of the gifts of minding and mindfulness. These gifts presuppose an ontological porosity to being, porosity of being. Artificial intelligence is wonder-less. It is the outcome of a calculating curiosity, itself the outcome of a curiosity that contracts wonder to what is more or less univocally determinable. The results of such artificial intelligence will be fixed as more and more perfect calculation, as we fall under the spell of the infinite expansion of the quantitative "more." Thus, we produce the *counterfeit infinity* of "mind," voided of true minding, which is *capax dei*.[1] We see the idolatry of this in the techno-scientistic-cybernetic totalization in the temptation to online totalitarianism. One might apply the point to robotic mimetics, not least regarding the quest to generate a counterfeit double of eros: eros without eros. If indeed our cybernetic-technicist-scientistic ethos is one of counterfeit doubles, of geometry without finesse, of calculation without wisdom, of "mind" without minding, should we be surprised to encounter those who advocate "religion without religion"?

I will refer to some other examples further below, but now I want to reiterate that my discussion as a whole is not a critique of curiosity in the sense of rejection of its potentially fruitful power. It is an exploration of modalities of wonder of which curiosity is one, one modality that is particularly zoned on determinability and our powers to determine the nature and intelligibility of being. There is a dynamic fluidity in the unfolding of our desire to know, and there are transformations that risk the deformation of that very fluid dynamism. I am exploring the enabling conditions of the desire to know itself. This is a kind of "step back" from the foreground of the desire to know, understood only as a kind of questioning curiosity. Questioning curiosity is an outcome rather than simply an original event as it is often taken to be. We tend not to raise the other question about what enables the desire to know at all. There is interpenetration between the modalities of wonder I distinguish, such that the matter is not a three-tiered cake, so to say, but a

1. Meaning: having a capacity for God.

fluid and dynamic forming, transforming, potentially deforming, unfolding of enabled powers of knowing in the very original porosity of being itself, in the original porosity of our own very being as given to be.

I

I am thankful to the shepherding editorial work of Paul Tyson, which is not to say that the voices included here are the voices of sheep. Paul does raise a hard-to-turn-down question at the beginning about how my work has been diversely received. The extremes: sympathetic thinking along with the work; antipathetic inability to enter into the work at all. It is important to distinguish between different communities of response, whether intellectual or professional, whether philosophical or theological, or other again. Some tell me I do make demands of readers, from whatever direction they come; others tell me, they come to find themselves relatively more at home, some earlier, some later. I do believe that if patient attention is given, at least at the outset, something of my voice and what I try to voice can be better heard. I have heard from those who hear my speaking voice that the reading becomes more accessible. Paul Tyson does draw attention to the ethos of instrumental thinking, materialistic metaphysics, the ethos of something like the dominion of serviceable disposability, and how these make hindrances to listening. Listening does not necessarily mean agreement or endorsement. Nevertheless, listening is necessary to hearing, and both needful for further judgment about the worthiness of what is being worded.

I do hold to a plurivocal idea of philosophy, and the orchestration of many voices is intended to do justice to our metaxological condition. This entails giving due consideration to the many models of being philosophical which we can find, philosophy being inseparable from its relation to its significant others.[2] I recall the fact that when I was younger, I did write for ten years or so a weekly column on philosophy for the *Cork Examiner* in Ireland, and believed in the venture of communicating to those not adept in a more professionalized sense.[3] I do not think I spoke down to such readers as I had, believing in readers not needing journalistic dumbing down. Aversion to what I then called "thought-bites" has deepened into intense diffidence for "twitter-thoughts," but my faith in human intelligence lives still.

Professional philosophers themselves are not always comfortable with the notion of plurivocity, even though many might well use a Platonic dialogue or two to introduce students to philosophy. This plurivocity does not always sing from the same page as the disciplining of philosophy according

---

2. Desmond, *Philosophy and Its Others*, ch. 1.

3. These provided first outings for the thoughts in *Philosophy and Its Others*.

to a carve-up of academic faculties. Professional philosophers do not always see the point of meeting a mark in terms of literary culture, and academic philosophy shows not a few signs of the flattening of prose into the banal. I pass over the challenge of religious or theological literacy. While in our culture one would anticipate a thinker being conversant with major scientific developments, if one is committed to a rich dialogue between philosophy and the poetic and the religious, it is perhaps inevitable that a mode of communication, attendant on plurivocity, will call forth resistance from different angles. It is a happy event for me to see the writers of these essays as not fitting into the category of resistor, even as they are questioners.

Much more could be said, but philosophy has become so professionalized that, beyond the division between analytic and continental philosophy, there is a tendency to multiplied silos of specialization where the busy management of research projects hides the loss of any sense of a more encompassing whole or community of mind. Even among the multiplying myriad of siloed scholarly monads, the fertility of the *metaxu* or between comes back to life again. In the *metaxu,* the space of communication between philosophy and its others is to be granted its promise, again and again. While one has to struggle against being contained in a silo of specialism, being conversant with the longer tradition(s) of philosophy provides one with an openness to a plurality of possibilities preventing specialized closure on oneself. I do not intend this in an antiquarian spirit. Quite the opposite, the thesaurus of the philosophical tradition can seed a fertile field within which new ventures in thought, entirely engaged with our current intellectual and spiritual situation, become needful and enabled.

This is the kind of engagement that the essays in this book diversely witness. I do not find antipathy to permeable boundaries between philosophy and its others; rather, I see suggestions for the reformation of the voicing of philosophy in its listening to and hearing of these other voices. Mindfulness becomes plurivocal in the porosity of the *metaxu*. In the *metaxu* one can meet resistance from the religious side, from the philosophical side, and indeed, from the scientific side, and from the practical side and the poetic. In these essays, I find practices of thinking that, on one hand, do not evidence disapproval or dismay that borders, said to be fixed, are crossed, and on the other hand, show approval that siloed separations do not have the last word. The conflict of the faculties (with a bow to Kant) need not return to keep us in our credentialed places.

Another "either/or" is not fitting: I do not fit easily the standard contrast of analytic and continental thought. I do not fit the first, though writing in English; I do not fit the second, because I resist the pieties of the post-metaphysicians. Both these sides have roots in anti-metaphysical, indeed anti-theological, tendencies. There is much in both echoing August Comte's scientistic deflation of a quasi-Hegelian trinitarian take on the fate of human history: religion/theology for children, metaphysics for adolescents, positive science for adults. We find Comtean children claiming majority in both the analytic and continental families. In one, a valuing of the crucial, sometimes preeminent importance of positive science reveals the direct echo of Comte. In the other, a widespread superannuation of the theological and the metaphysical, echoes pater Comte from a different direction. There is a pluralism of scientisms—perhaps a reflection of the plurivocity of being, now expressed in the multiplication of candidates to being the one science, needful for all things present and future. We diversely reach out to take hold of the *potestas clavium*. I believe a mindfulness more metaxologically ecumenical is needed. And this, not only in the practicing of philosophical mindfulness, but also in the companioning space where the others of philosophy offer, and are to be allowed offer, their distinctive voice.

It is clear we need more than binary thinking: not just univocal versus equivocal; binary versus nonbinary; determinate versus indeterminate; fixed versus fluid; cybernetic versus promiscuous. We need dialectical thinking and trans-dialectical. I think the fact that I write in English yet in a manner between system and poetics is not irrelevant to the keeping alive of wonder, and to the worry about different scientisms. Is the poetic only ordinary language put to more than ordinary purposes? Or the invitation of something more original and a response? And our philosophical response? I have wondered if English-speaking philosophers have lived up to the spiritual seriousness of their poets, such as Shakespeare or Wordsworth. One need not be English to write in English and experience the challenge to philosophical wording posed by great poets. When I recall the spiritual seriousness of great poets, the fact that scientisms flourish in the widespread belief that "science" is the only avenue to being and truth (to say nothing of goodness and beauty), I am made to fear that Ockham's razor is now a cut-throat to spiritual seriousness. The cut throats of metaphysics and theology are allowed only to gurgle a welcome for their own demise. Being metaxologically mindful is not a matter of reductive objectivism or relativistic subjectivism. It seeks to remain true to the originating sources in porosity and astonishment, manifested significantly in a plurality of ways of being true. Plurivocity is a provocation and not a problem: provocation as a calling forth (*pro-vocare*) and one worth trying to meet, especially in making permeable again silos of significance, now shutting off or shutting out. Thinking becomes newly a wording that communicates a betweening.

I have been a defender and practitioner of metaphysics for many years, and this in a time when the temper of things has been very anti-metaphysical, and anti-theological. There are many sides to the story, but I only note this: my sense that some readers, while drawn to the concept of the between, early on have an anticipation that the exploration must inevitably bring them to the question of God, and that is a question that they shun, hence further engagement is curtailed. I have this image of walking along the street, and someone is coming towards me and the possibility of our meeting is real, and then the one approaching, anticipating that meeting and anticipating something in it not sought or welcomed, the God who might companion all ways of walking, crosses the road and passes by on the other side. Perhaps it is an exaggeration, but I sometimes think such a passerby smells God from my work, and this does not carry the perfumes of fragrant divinity for them but rather reeks of regression to now-overcome metaphysics or past-and-gone ontotheologies. One has to have a nose for such things.

II

I turn now to **Spike Bucklow's "Preparing to Paint the Virgin's Robe."** I note we begin here with painting the Virgin's robe, and will end later with another beginning, namely with Isidoros Katsos's reflection on an icon of Christ's baptism. Among the many things I find appealing in this reflection was the reserved statement that contained multitudes. I am thinking of the opening sentences referring to a kind of release that can come once having peaked professionally: one is then not quite constrained to package ones "projects" in terms of normalized routines of (seemingly) addressing curiosity. One is released to ventures having to do with beauty that transport us beyond obvious utility, in this case, in trying in understanding earlier art to bring back to mind what seems to have vanished. And this, in his case, by an enlargement of mind, and not a contraction of (scientific) wonder. He finds himself endorsing the historian Barthold Niebuhr suggestion that "He who calls what has vanished back into being enjoys a bliss like that of creating."

These remarks set the tone in many ways, which I recognize also from his inspiring book, *The Alchemy of Paint.*[4] This is a marvellous book about a world of wonder(s), wooing us as a lover woos, into the mysteries of colors, the elements, the abiding power of gold, and more. There too the Virgin's robe makes an appearance, divinely dazzling. We learn of a world, and relevant to the current reflection, where significances—practical, scientific, magical, symbolic, alchemical—were communicated from adept practitioner to learning apprentice, in a line of inheritance in which the directing suggestion or gesture of the adept, more than some univocally fixed method, signalled the way a follower might pass and be passed

4. Bucklow, *The Alchemy of Paint: Art, Science and Secrets from the Middle Ages.*

192 ASTONISHMENT AND SCIENCE   PART THREE

along astonishing secrets. There might be recipes, but as with all recipes, the blessings of the sprinkling hand can be all important to the savor of the outcome. This is a world where things word themselves, word themselves because first worded into being. Such a world is closer to the all-in-all of the original porosity of being, though this is no way compromises the precise particularities in permeable intercommunication in that world of wonder. We have scientistically univocalized the particulars, and fallen into forgetfulness of the overdeterminacy of being in the mixture of the porosity. Such a world loses its savor.

There is no denial of determinacies, but a deeper exploration of what certain determinacies reveal shows what univocal determinacy alone cannot fully reveal. In speaking of release after professional peaking, and offering us these reflection, I see something of a truer freedom to explore and consider things anew, beyond the sometimes extraneous (institutional) consideration that contract the public field of one's engagements, and perhaps publications. Spike shows how the established practices of modern art history can well be impressive in establishing certain determinacies about an artwork, and can be revealing of things that otherwise pass notice. And yet this revealing can well be a concealing: a fixed determinacy can hide something either more indeterminate or overdeterminate, in the instances cited here, revealingly overdeterminate.

Take his speaking of the ambition to identify a Raphael on the basis of scientific analysis as arguably an example of scientism. Perhaps one might say this only if the claim is forwarded as the real truth of the painting. The analysis may tell us something determinately true but in thus being true it may box in the overdeterminacy untruly, and hence in being true it is being untrue. The way to scientistic univocity leads in such equivocal directions.

I thought Spike hit more than one nail on the head (too violent a phrase perhaps in its peculiar determinacy) when raising great questions about the process of grinding and cleaning azurite to moderate and achieve the brilliance of the color. The questions are great in their seeming simpleness. First, in the process, what is a "light touch"? Next, how to "stir"? Then, how long are we to let the water "stand"? To take the latter: a judgment call is needed, since letting it stand longer gives you more blue, though the mixture is paler; while a briefer period of letting stand gives a deeper blue, though in a smaller amount. Our attention is being drawn not only to nuances of brilliant color but to practices of finesse, even though the steps taken cannot be exactly pinned down with univocal precision. His account of the process grants its indebtedness to modern science and yet there is a poetics beyond science, a poetics itself exiled from modern science at its inception.

This reflection reminds us how we might oscillate between a tempta-tion to scientism and yet find a pull back from the temptation, in the very refinement of attention that is more faithful to occluded nuances. More, nuances themselves come to the fore in a fidelity not untrue to the desire for the true that we find in true science. A more released mindfulness and a more finessed attentiveness can be allowed to come to the fore.

The brilliant colors of this too-forgotten world of wonders makes us wonder now about what we have, and what we have lost, and how current practices themselves call for new or renewed formation. And perhaps this applies also to the practice of thinking. Something that comes to mind is Aristotle talking of the life of *theōria* as a *bios xenikos*: a strange life, perhaps a life startled by the surprise of the strange, a startlement itself strangely intimate, and thus not merely estranging. Why theorize, Anaxagoras asked? Anaxagoras answered: To behold. I find here in this reflection, and the prac-tice on which Spike meditates, an enjoyment in beholding.

I will speak again of "beholding of" and "beholding from." We do not have to erect any binary duality between the exactitudes of scientific/techni-cal thinking and the nuanced "beholding from" of art history. A more fluid hithering and thithering is not impossible: and yet in the more released "be-holding from," the univocal exactitudes serve the saturated equivocities of finesse. One's soul and its diverse powers must already have come to a point of mindfulness that has ceased to be siloed: ceased to be siloed within, rela-tive its different powers, and ceased to be siloed without, relative to our ever open engagement with what is other. Spike Bucklow's reflection suggests a marriage of determinacy and more than determinacy. There is more exact curiosity in which the originating wonder continues to germinate and not commit scientistic suicide. It seems that univocity and a feel for equivocity can be companions, just as can geometry and finesse. Too often we jux-tapose objective versus subjective, and instead of finding a companioning relation which might be both, or neither, we bog down in sterile dualisms. Does one have to attain superannuation to achieve this released poise? Or is it sometimes so that the time of superannuation is the time of career attri-tion of the originating wonder that lured one, when young to it, to the thing one's mind then loved?

We find it hard to find poise between different modalities of knowing: knowing that, knowing what, knowing how, knowing who, knowing not knowing, not knowing. Knowing what and how often dominate. But one can come to know that, in practice, knowing that, knowing who, know-ing not knowing, not knowing can well be in a fertile synthesis. This can lift the endeavor of seeking to a new celebrating level in which curiosity is redeemed in a "beholding from" that is rejuvenated even as it ages.

Our ventures in technical knowing too often are still in a fugue state from originating astonishment. I think from my own experience of students attracted to philosophy. At first, something strikes them and they are awakened and opened up by unanticipated surprise. Later they will come to the conclusion that they have to have a "project" to be involved in the business of academic philosophy. Being philosophical mutates into becoming a scholar manager, first as an assistant, then as supervisor. I do not see quite the learner and the adept in this model of scholar manager. Has the philosophical Parnassus been climbed? Where might one apply for support for such a climbing project?

The invocation of the medieval workshop brings to mind a differently accented ethos of wonder. Asking of the apprentice a habituation to the finesse of the master or adept (presuming there to be one), it can lead him or her to gain a feel for nuance, in intimate porosity between us and the things. While this might seem only indeterminate, it is more than determinate, it is overdeterminate. There are plurivocal betweenings that are much more than inputs and outputs. No doubt there was the equivalent of the latter: a workshop had to pay its way in letting its artisans make a living. Making a living: there are crafts calling on things that have survived the test of time, a test sometimes full of caprice and bad fortune, hence itself an uncertain carrier of achieved excellences. There is a message here too for the participatory character of philosophy: a workshop of philosophy finds work beyond work, and can open again onto a refreshing oasis in the desert of serviceable disposability. I see a connection with reflections on Stengers's idea of "slow" science, spoken of by Simona Kotva, as well as a question for the sacred art of icon-making central to the contribution of Isidoros Katsos.

III

Steven Knepper offers us a reflection on "**Cultivating Wonder**." It is a pleasure to read this reflection, indeed to savor it. I take it as written in a kind of companioning spirit, and I offer a response in a like companioning manner. Steven Knepper has written marvelously well elsewhere about my own work in relation to literature. It is not insignificant that the title of his book is *Wonder Strikes*.[5] It seems to me that wonder strikes again in this marvelous essay. There is a redoubling of wonder in the way Steven is struck by the wonder of his daughter on her beholding the things of nature. In beholding

---

5. Knepper, *Wonder Strikes: Approaching Aesthetics and Literature with William Desmond.*

the way wonder passes to his daughter, the father finds a second wonder passed to him. "Passing from" is as elemental as "passing to."

I have made references to "beholding from" above, and I take the phrase from Wordsworth in order to describe a reversal in our normal sense of perception as going from us towards the things, one rather highlighting something being communicated from things to us. "Beholding of" is subtended and exceeded by "beholding from." Something comes to us from beyond ourselves and we are opened, and original astonishment or wonder is just that happening of being opened by the otherness of being and beings. Steven Knepper situates his reflections very personally in relation to his own upbringing, and there is an idiotic side to this in relation to an intimate singularity: idiotic not as a tale told by an idiot signifying nothing but a tale telling something asymmetrically beyond nothing, in its incontrovertible, though often not noted, being there.

I can concur strongly with his own experience of nature, and again the echoing of Wordsworth is not without its grace notes, especially resounding of those earlier formative years when we are more porous to being's massive thereness, its being too much in itself to us and for us. This is more truly phenomenological in a realistic sense rather than transcendentalist in an idealist sense. One might reflect here how the Romantic generation tended to be perhaps too much transcendental in the transcendental idealist sense, where the otherness receives its meaning more from us than we receive from it. There is something of this in Wordsworth when he writes, "of all the mighty world / Of eye, and ear,—both what they half create, / And what perceive." I asked: is this entirely true to the "from" in "beholding from." Coleridge embodies something of the condition perhaps more in *Dejection an Ode*. Coleridge: "O Lady! we receive but what we give, / And in our life alone does Nature live: / Ours is her wedding garment, ours her shroud!" As I put it: If we receive only what we give, there is no "beholding from." If there is "beholding from," we receive before we give. "Beholding from" asks our attendance on the things that *come to mind*. I recall these matters only because I sense that Steven himself, and in relation to his child, deeply understands the priority of the receiving.

The resonance of wonder striking is heightened more in Steven Knepper's *giving on* of what he has already received. It is not beside the point that it is with reference to his own child that this giving on seems raised to a second power, so to say. The girlchild is father to the man, to adapt and turn around the old phrase. Surprise in the eyes of the other awakens our surprise in the beloved other and in what surprises the other. There is a metaxological contagion, so to say. In Steven this is not an ode to dejection in which the deflated poet longs for the inflation to come; for indeed coming precedes giving, precedes giving out and giving on. The strike of wonder is a primal perforation of any being closed in on itself, experienced later perhaps as an alarming wounding of the shut-down soul.

Steven is also surely on the mark when he tells us that the child needs the nurturing of an adult to help sustain and encourage and, sometimes in the lucky, even providential case, to direct our wonder to what itself is wonderful. Again metaxological intertwining is both unavoidable and needful. There is something of agapeics in this, since the generosity of love gives for nothing beyond the giving. And after all, being struck by wonder is not unlike being stricken by a mysterious love.

In a way it is analogous to the awakening of the impulse to word things, our power to speak: this would not be properly called forth did not another speak to us. We speak of the very young child as *infans*, without language, but being without language is not necessarily without the capacity to be opened to the things of the world and to find our openness itself overtaken by words expressing admiration and surprise. Gosh! Wow!: the words are short ejaculations in the sense in which short prayers are utterances of amazement and indeed love in the face of a gift given to us. All of this is there in Steven's essay. I should say too that there are many children for us intellectuals: the child of Heraclitus, the child of Jesus, the child of Nietzsche's Zarathustra. I just note that Zarathustra's child seems to have no father or parents: self-born, a self-propelling wheel. From where does the originating strike of wonder come? One might say Heraclitus has the answer in saying that polemos is the Pater of all things, and yes while there is a Pater, and indeed Zeus is himself a *pais*, a son, is there a more original father to polemos as father, making this latter father also a child. I think the strike of wonder, even let it be it a lightning bolt that breaks open and lightens up the sky, suggests the more original source of the sky and the light in the sublunary world. None of this can be understood scientistically, indeed perhaps scientifically, for we are being struck by wonder now into the Father of heaven and earth and, awaiting the returning wonderful child of creation, of divinely endowed time.

While personal, Steven's recounting is not merely personal, in the manner in which his daughter's sense of the selving of things comes before our gaze. Selving: this is a word that Hopkins uses to describe not only human selving but all things as selving. I think this is certainly nested in his daughter's experience as well as his own, and lends support to his claim that poetry, literature can itself be a companioning sustainer of originating wonder, and not simply confined to the specialized sense, because it defines the human being as a poetic being in the midst of beings. It is no wonder that we do often appeal to the poets to remind us of what in the familiarized world we take too familiarly. He does report how his early world was also saturated by religious significance. This is important, for this sense is sustaining of our orientation to the mystery of given being, relative to what gives it, relative to what gives it to be, and relative to what it gives yet to be: origin and end, both as graced by something not determinable in finite terms alone. Being horrified: this is not excluded at all; far from it, but it is a relevant question as to whether horror itself is a mutation of astonishment in light of an otherness, threatening and menacing and even malicious, more than our determination can manage or comprehend. The monstrous arouses wonder in the space of the sacred as much as does the beautiful and the holy.

I cannot forbear another child germane to conversion, away from and beyond thinking as negativity and indeed beyond scientism as a form of contracted wonder. I am thinking of the mysterious child whom Augustine heard in the garden and urging Augustine: *tolle lege, tolle lege*.[6] The child in the garden. Reading: *tolle* as engendering a saving toleration not simply in the sense of carrying a burden, but more of a charge, perhaps a cross. Reading can itself be a child engendered by "beholding from," born of original wonder; generating from one to the other the word that passes in the between of communication, even to the last syllable of recorded time. Sacred syllables, one holds in hope, not desacralized, even desecrating syllables that signify nothing. The child in the garden asks our reading that will bring us back to the garden and read with true finesse, opening anew the sacred signs of things in the promised garden of the world, the new creation. (These last remarks remind me of the world depicted by Spike Bucklow in his book.)

There are strikes of wonder where we become again as a child, and we live again from our original opening to being in the elemental given porosity. Becoming a child again is not becoming childish. Perhaps the most extraordinary, and astonishing address of intimacy of the child is when the Divine Son calls, and calls on, the Father as Abba. I suspect an aversion to being a child in the scientistic impulse. I am apprehensive about onslaughts against the father (the "patriarchy"), indeed against being a mother. (Consult Aldous Huxley's *Brave New World* where the family and mothers are treated as horrors.) With scientism, whether Comtean or otherwise, being a child might be a necessary beginning but the development of science makes it something necessary to be transcended. Being religious is the childhood of the human race, and something to be surpassed. Scientism must allow nothing childlike, for after all has not humanity "come of age." Is there here a secret fear, if not contempt for childhood?

6. Meaning: "Take up and read, take up and read."

Of course, there is a coming of age, and sometimes this is the second childhood of dotage. Whether in the beginning or in the end, the advance beyond or retreat from the disturbing surprise of originating wonder can be overtaken by intrusive curiosity. Nothing of the wonderful love communicated between a parent and child can be captured by such intrusive curiosity. Childlike, when old, we can find ourselves again the porosity of being, with an astonishment not produced by a project, but opening us again as it once did, and there is new joy in the light. Wonder as astonishment is more alive as the child lives this primal and elemental opening. The parent can relive it in the wonder of the new child (Simon Oliver too has something to say about this). Scientism would kill the child of astonishment in the womb of wonder and put in its place a changeling of loveless curiosity.

IV

I found **Simone Kotva**'s discussion of **"The Astonishment of Philosophy"** very illuminating on the convergences and contrasts between my thought and that of Isabelle Stengers. I hear Simone Kotva as perhaps a kind of Augustinian child in the garden of thought urging me: *Tolle lege, tolle lege* . . . . I am grateful to have Stengers's recent work brought to my attention and to be nudged in the direction of coming to understand her more adequately. As recounted here, among other things, her work is concerned with "slow" science, and I find it interesting that a word like "slow" can carry opposite connotations with respect to intelligence. We speak of a "slow" child, and we mean someone to whom understanding does not come easily. This is not the meaning of Stengers, of course. Her meaning points us towards the more marinated mindfulness of an elongated time, opening an alert attentiveness, coupled too with a different organization and practice of scientific investigation.

That the same word can carry us in different directions has significance for the word "curiosity." In my thought I want to allow for the extremes: curiosity as fostering alert attentiveness that is not *merely a foster child* of original astonishment; curiosity as impatient intrusion into the resistant mystery of things. The first fostering is more than just being a foster child of original astonishment; it is truly a child, and hence genuinely fostered. By contrast, the second curiosity refuses to be fostered by anything at all and even takes itself as self-originating in its exhilarating hurry of claiming to be the measure of things. I would say true "slow" science is fostered on the original porosity of being, on our original porosity to being, in which our *passio essendi* marries receptivity with the (admittedly sometimes headlong, sometimes headstrong) desire to know. That desire can be sometimes headlong, overtaken by an endeavor to know; it can also be paradoxically head-strong in its patience to being as it is.

In the overlap of our concerns with "slow" science, the tasteless glass-house fruits of scientism are not what we seek to savor. That said, I wonder if "slow" has quite the same contextually defined character as a modality of wonder as this does for me. I am thinking of another crucial word mentioned by Simone Kotva, namely, "vocation." First, she is right to say that I am not intending to pit science against metaphysics. I think no such pitting is possible if we understand metaphysics as I do, as dealing with the fundamental senses of being, and the enabling conditions of our being and our being mindful that allow us to come to some knowing of being. Metaphysics can offer thought on an ontological archaeology of the endowment of the desire to know, as well as suggest the promise of a theological teleology. Science itself embodies for me the desire to know, tempted to become a drive to know with as mathematically precise determination as possible with respect to univocity. It is not fitting to fit all of reality in accord with this drive.

There are different univocities, I do think: what univocity means in common sense is not the same as what it means in mathematics; and there are other incarnations of univocity I cannot here treat. [7] While wonder as astonishment is an enabling source more primal than determinate curiosity, I see wonder as both astonishment and curiosity each as at work in the scientific enterprise. Nevertheless, the tilt there of the formation of wonder is less lingering on mysterious origins and (re)sources as driven forward to make as much determinate intelligible sense as possible. At stake is a venture of explorative understanding seeking to be true to the intelligibility at work in being. Science is an enterprise turned to being, but it can forget the enabling in turning its face towards given being as other. I am thinking of natural science considered as object-process oriented. It is not concerned with the knowing of knowing itself: with knowing itself and with itself as knowing. Philosophy is an enterprise in a different space dealing with these latter. If we are doing philosophy of science, this is not itself science. I see transcendental philosophy coming from the Kantian line of inheritance as seeking a kind of transcendental univocity in the subject, as putatively the Copernican sun in immanence. This transcendental subject does not receive the shine of the sun; it claims itself to be a sun that shines intelligibility, one, with all due qualifications, of its own making. I think the sun of Plato is truer to our being given into the space of truthfulness. If it is truly the sun that shines, Plato is more Copernican than Kant.

7. See Desmond, *Being and the Between*, ch. 2.

That aside, philosophy is not just self-knowing—it is an attempted mindfulness of what ontologically, what in being, allows knowing to know itself and what is other. I sense these concerns are not foregrounded in Stengers's work as presented here. I find more a presentiment of something lacking in the contemporary practices of science that tends to close the doing of science down, at least with respect to the other questions that are in another dimension. This applies to the question of self-knowing, as well as the knowing of being as other. This is why I connect a contraction of curiosity very much with scientism, despite the surface appearance of curiosity being open to all possibility. Its openness is qualitatively inflected relative to the receiving porosity of original wonder, and perhaps the renewed and celebrating porosity of human wonder, at the limits of its own achievement of comprehension.

I am very much in tune with the "hesitating" stress in "slow" science. The word calls for a correlative attunement (another theme I merely mention) one would recommend to be present more than it often is. I myself would hesitate to channel "vocation" in the direction of the "unhesitant." This perhaps has something to do with words, but in my lexicon the "vocational" nature of philosophical thinking makes it a member of the same family as the "vocation" of the priest and the poet. Stengers might not endorse thus mentioning the monastery. I think the non-slow science has more to do with managers than monks. The scholar or scientist as a manager of projects is not called forth in reverence for the truly real. This manager is happy having a portfolio of funding that apparently allows one to cast a net of anticipated intelligibility over a domain of research. The manager is trammelling up the future of the "perhaps." And perhaps thus the manager is retarding (another kind of "slowing") the waywardness of the surprise we have in meeting a revelation not dreamt of in the stated terms of the project.

Interesting to me is the reflection on the displacement of the "monastic" from the monastery to scientific study in early modernity. It is notable that this is found in cultures themselves hostile to "monkery." ("Monkery": a word used by Mrs Flood in appalled recoil to Hazel Motes's penances in Flannery O Connor's *Wise Blood*; but consider the diatribe against the "monkish" of the genial Hume.) Such a displacement risks generating a counterfeit double of the monastic vocation: a calling without being called, a vocation without vocation. Why? Because curiosity in time, buoyed up with its "successes," yields more and more to the scientistic temptation and turns wonder away from slow contemplative mindfulness under God. We become successful because we make ourselves superficial.

In some educational systems "vocational" is correlated with utility and in older language the *artes serviles*; while "liberal" in a humanistic sense is

associated with a freeing beyond utility alone: artwork beyond just work—
because at work in the *artes liberates* is something more than (instrumental)
work. The work is its own justification, in a way; and the work that is thus
its own justification is not work. Paradoxically, such work might be slow
with eternity even though it be fast with time. It is not punitive for a fall
but a free enjoyment of endowed power. That sense of a freer pedagogy
is severely crimped in our current economy of scientific projection. Simone Kotva is very right to remind us of the (sometimes forced) servility of
younger manager-scientists/scholars in the regime of projects. I have myself
seen the changes I predicted decades ago due to the regime of projects: no
longer mindfulness on the look-out for what may come to show itself, even
despite the project(ion); no longer a love for something intriguing, incho-
ately sensed at the outset, but sought with ardor beyond utility. I know there
are arranged marriages that can turn into loving relations. But how many
arranged marriages in the regime of projects express love or foster love?
Simone cites provocatively the analogy with grooming and paedophilia. But
here the child of wonder is abused. For where is there true *philia* of the child
here? Far from fidelity to vocation, or *philia* of *sophia*, one sees subjection
of research students in the regime of projects. Free art is remade as servile
work. I see the betrayal of the freer mindfulness that "vocation" carries for
me.

Simone Kotva's remarks are very much to the point concerning how
this freer mindfulness makes one seem "slower than their peers, less efficient and less productive"—slow as more considering and yielding more
considerate views. She is right to remind us of the "marriage" of scientific
institutions and industry in Western-led global capitalism. Instead of a more
searching openness, there is the fact that only certain kinds of research are
funded, while others receive no support. On this global scale it is very difficult even to raise the issue of the counterfeiting of wonder in scientistic
curiosity. Appositely she says: "In the capture of curiosity by the scientific
institution, curiosity is refigured in ways that are structurally analogous
to the repurposing of environments, societies, artefacts, and organisms by
capitalist economies. Rather than perceived as gift, curiosity is seen as resource." In truth, gift subtends use. Can one even hold to a distinction of
basic and applied science when the paymasters of projects have power over
the first as well as the second? Original wonder risks being trapped in a
Babylonian captivity that would circle around itself on a global scale. There
is this too: if the global crowd is rushing ahead to who knows where, and if
one wants to slow down and look, one might give the impression of being an
impediment to forward motion, or even a patsy of regression. Yet the pause
of the slow would be the truer wondering.

I do think Simone is right to remind us that in addition to not pitting metaphysics against science, the matter is also one of speaking of a "zone of mindful practice" that would transform the theology-science debate. I am put in mind not only of Pierre Hadot's recalling us to ancient philosophy as practices of thought marking ways of life, but also of Ryan Duns's efforts to see the matter of a metaxological approach in terms of a variation on spiritual exercises (with a bow also to Ignatius of Loyola).[8] Mindful practice comes from endowed porosity and our vocational fidelity to it—the endowed porosity is ontologically constituent of our being.

Towards the end of her reflections, I was struck by her remarks about my reflections as "the philosophy of astonishment *par excellence*," and made to wonder about how I "know" about astonishment, since there is something here exceeding determination. A difficult and important question, at least as I am posing it here. I do seem to "know" something about astonishment, and if so, is this a kind of performative contradiction in that in calling determinacy into question one must call on determinacy? My point is never to deny the constitutive unavoidability in being of determinacy, and the indispensability of curiosity in knowing such determinacy. Nevertheless, in the midst of determinacies the more than determinate manifests itself. This is part of what is at stake in the hyperboles of being.[9]

There is also a "not-knowing" that goes with such "knowing." There is a "curious" kind of knowing that is in the business of warning against curiosity turned to counterfeit knowing. The negative side of this is not negativity in either a Hegelian or other sense. I think more of Aquinas: *videtur mihi ut palea*.[10] I take this to be especially true when said with respect to cognitive pretentiousness vis-a-vis determining divinity. And yet there is a knowing in this not-knowing and in a not entirely negative sense. The knowing is not the knowing of a determinate somewhat—it is more importantly a *waking* of knowing to astonishment that it is knowing at all, and an archeological venture into what makes waking up at all in knowing possible. Being mindful of this is bound up with a mindful life or way of life. There is an ethical and religious call in this. And the matter is not simply transcendental in the Kantian sense but ontological, and indeed metaphysical in terms of what one comes to know about knowing. There is no derogation from knowing, but knowing, like the desire to know, is made possible by a porosity that does not know itself in ether a determinate or self-determining sense. Philosophy itself is not self-originating. This is the idiotic dimension and enters

8. Duns, *Spiritual Exercises for a Secular Age.*

9. Desmond, *God and the Between*, ch. 6.

10. Meaning: "it seems to me as though straw."

intimately in the happening of what I call idiot wisdom. This is not dreamt of but rather fled from in the practices of the determinate sciences, even if we reform them in the light of a metaxological metanoetics.

V

In **"Astonishment and the Social Sciences," Paul Tyson** offers a challenging and reconstructive exploration that makes me think that social scientists seem often to inhabit a world that is not the full *human* world, and that a metaxological reform of their practices might bring back something of this fuller reality, bringing them back to themselves too. One wonders how they can treat human beings as if they were not creatures of wonder and astonishment; and how the dearth of astonishment in their own approaches, too, seems to match their contraction of the human.

Paul outlines two broad trajectories in contemporary social sciences, the positivist, on the one hand, and the other hand the tradition stressing *Verstehende Sociologie* (interpretive sociology). Among the first he includes those indebted to French theorists like Comte and Saint-Simon, and those more pragmatically, materialistically, and analytically oriented in recent Anglo-American thinking. They seek measurements and offer statistical analyses of correlations between social conditions and social behaviors and attitudes. The other group study socially situated meanings, suspicious of statistically measurable causal dynamics; there is more at play in human society; there is meaning beyond any machine, culturally and historically shaped and communally embodied. In practice, positivist and interpretive sociological trajectories often have quite a bit to do with each other. And this is not necessarily a bad thing. Paul offers a very intricate discussion of significant trends in the social sciences that he holds would benefit from a reform in light of some key notions I have proposed. I cannot say otherwise but that I agree. Still, though the modalities of mindfulness and particular formations of curiosity in the social sciences does threaten one with a sinking feeling that one is as in an epistemic dream when one knows one must move but one's legs will not answer the call and one faces the horrifying prospect of being stuck in place. Stuck in place while all around one busyness is everywhere, and business is as usual. And yet it is all of that business that wakens the desire for a different knowing and a different refreshing of the sense(s) of being at all.

Paul reminds us of moves towards a conscious synthesis of positivist and interpretive approaches to sociology in terms of what some theorists have called "reflexive sociology." While detailed quantitative and statistical analysis of social phenomena will be done, sociology will be reflexive in trying to be aware of its own interpretive commitments, not all of which are to the fore of self-consciousness in more positivistic inclined approaches. My work is suggestive of possibilities that sociology, however, cannot be characterized as either positivist or interpretive. Among other things, Paul draws attention to the dialectical relation of fact and interpretation. There is also the fact that a metaphysical openness is "glaringly absent" from the positivist and the interpretive traditions in most contemporary sociology scholarship.

Worth remembering is that the menu of possibilities Paul Tyson outlines all have their roots in sets of philosophical presuppositions or orientations, be they covert or overt. The philosophical self-interpretation of sociology shapes, to a greater or lesser degree, the disciplinary practice of sociology. Philosophical shifts, not devoid of theological possibilities, produce different practices, and this is at the core of Paul Tyson's reflection. My bad dream of paralysis need not be the last dream. I am not entirely bereft of hope when I read Paul's invocation of prophetic fire in light of Kierkegaard's theological sociology.[11]

In responding to this discussion, the figure of Auguste Comte comes to my mind again because of his trinity of religion, metaphysics, and positive science. Paul mentions metaphysics and positive science but in his thought the religious is not to be neglected. Comte might sometimes be mentioned in introductions to sociology as one of its founding fathers, now well superseded by more precise, more quantitative social methods and data, more expertly, more methodically gathered, leaving to the founding father more of the big idea(s) than the detailed studies, more humble in their underlaboring, and more cumulatively significant in expanding the empire of data about the social human. It is true that he does remind us of currents in social thought that have come to understand how "theory-laden" the gathering of social data can be, and something to be taken into consideration and perhaps guarded against as unconsciously biasing us. And yet one wonders to what extend the big idea(s) of the father, Comte, are out of play, while his more "successful" successors gather quantitatively the social harvest.

My own sense is that the triadic phasing of human childhood, development, and maturation continues to hold a sway, even in thinkers who would smirk at the exorbitant totalizings of Comte's proposals and projects. I believe that the modes of wonder are, to a degree, mirrored in Comte's historical scheme of religion, metaphysics, positive science. Religion for the indeterminate beginning of childhood, metaphysics for the rationalistic

---

11. Tyson, *Kierkegaard's Theological Sociology: Prophetic Fire for the Present Age.*

determinacies of the abstraction-loving adolescent, positive science for the scientifically mature human, intellectually come of true age, both theoretically and practically, and redeeming human promises of the consummate prospect of being "all in all." What drives the move from the first to the second to the third, except a will to "scientific" determinacy, itself tempted with a will to self-determining totalizing in an exclusive human sense, both theoretical and practical?

There is a sense in which Comte's scheme comes to life again and again. I think even of Heidegger as an exotic kind of Comtean, in seeming to frame the entirety of the Western tradition as moving from myth through metaphysics to cybernetics. It is all Comtean ontotheology, so to say. There is an echo of the later Comte too in the resurrected reverence for pagan myth prior to the fall into Platonic erring with *eidos*. Comte did come to see the need for something resembling a new animism in the positive/positivist age. Positivistic curiosity creates its double of the religious reverence previously overcome, and indeed in Comte's case, a kind of counterfeit double of papal Catholicism within the institutional and liturgical forms of the Positive Totality.

I know that Paul Tyson gives us a more detailed picture of current sociological practices that would shrink from being identified with any such form of positivism. But I think if we undertake a metaxological "step back," something of the Comtean triadic phasing of world-history would not be lacking in insinuating influence. No one could call Heidegger a positivist in the classical sense, and yet if we "step back" from Heidegger, it is striking how Comtean the history of *Sein* seems to be—at least within the frame of ontotheology, now totalizing itself in the cybernetic worldview. One wonders if there are closet Comteans also to be found in those following Heidegger, naming themselves as postmodern and claiming to be post-metaphysical. One might think of the "return of religion," but in the equivocal form of "religion without religion." If one were asked to put an identity on this, one might have to put it charitably as perhaps a first cousin to the counterfeit double.

I find it instructive to return to the question whether sociologists do suffer from a dearth of astonishment, especially in the more ontological/metaphysical register dealing with the human being, and its flourishing or languishing in different societies of togetherness. Would we not have to contend with a double dearth: dearth on the part of the sociological inquirer; dearth attributed to the "objects" (that is, social "subjects") of study. A dearth on the first side almost inevitably will find in the "object" what is not to be found in itself as inquiring "subject." The dearth will be self-perpetuating, and will seem self-authenticating, given that no terms beyond this self-circling of dearth will enter for consideration. If the investigation, suffering from dearth of astonishment, goes to work, it will perhaps seek to inoculate itself against the surprise of otherness. Of course, in the nature of the case, it cannot do this, in the end, though in a long process of self-incurred stultification it may well congratulate itself on now at last attaining to more authentically scientific status. Void will call to void, but without the abyssal significance of the archaic *Abyssus ad abyssum invocit*.

If such busy and (to all appearances) successful curiosity is asked to deal with counterfeit doubles, how could it manage at all, given that its formation of curiosity is always called back to and anchored in a counterfeit form of wonder? I am thinking of social infections like "wokism" and the counterfeiting of enlightenment by which it lives. How apply itself to woke phenomena if it and they are both counterfeit doubles? I think of counterfeit astonishment in cybernetic spectacles; of counterfeit outrages mimicking righteous wrath; of counterfeit compassion doubling for agapeic generosity; of curious research and counterfeit commentary that is the cybernetic counterpart to the prisoners in the cave getting a hold on the flickering of the images before them, entirely sleeping to their own captive character. How are we to be released from a thinking about things, a thinking about ourselves, that are themselves the outcome of the counterfeiting? In the encompassing circle of counterfeit doubles, we cannot really see what is going on. Our seeing is so shaped and directed that we cannot see otherwise than in terms of the counterfeits.

Paul gives me credit for taking issue with the Kantian inheritance with regard to transcendence and I think he is right to bring in philosophical considerations that make our openness to transcendence as other a perplexity asking more of us than Kant offers us. I often think of Kant as pacifying our metaphysical perplexity on a *transcendental Alcatraz*. This is the way Kant describes truth: an island surrounded by mist banks of illusion, and shape-shifting forms that induce deception such as may seduce us to trust as true what vanishes in the mist. It is the spirit of anxiety and suspicion, of meagerness and defensiveness, that sets us to countering untruth rather than encountering truth, with the enclosed space of the island contrasted to the vast unlimitedness of the surrounding deceptiveness.

I cannot but think this is a miserable little island of truth, and Kant sometimes seems more afraid to get his feet wet than to follow his own encouragement: *sapere aude!*[12] In these words, a purported motto of Enlightenment, he claims to champion audacity, yet he remains safely on the shore and will not venture out beyond. Perhaps his caution is understandable,

---

12. Meaning: "Dare to be wise!" or "Dare to know!"

given the uncertainty as to whether there is anything out there about which Kant could claim certain and certified cognition. There certainly is nothing like the audacity of the Irish St. Brendan, who in a frail vessel ventured out into the watery limitlessness. Kant was more content to continue doing his mechanically clocked walk around his little made-to-human-measure island. This is more like a prison island than an island of truth. Kant is not out in the world—he is transcendentally framing it in representations—the frame of subjectivity, looking out its window from the island surrounded by mist banks. Is this frame a transcendental prison, the transcendental Alcatraz of subjectivity—every room with a view? And one will never leave the island and live. It hardly helps to now appeal to our cybernetic culture of screens: the transcendental Alcatraz of the iPhone age. We do not see things (in themselves), we see only appearances (for us).

VI

Andrew Davison's **"Curiosity, Perplexity, and Astonishment in the Natural Sciences"** is a very perceptive joining of Kuhn and myself, offering a well-handled analogy of our respective concerns, in light of the different modalities of wonder. In it I found myself instructed about Kuhn, and I hope others will find what I say instructive in a fruitful way. I find helpful the suggestion that my category of curiosity might be correlated with Kuhn's "normal science," my sense of perplexity as relevant to periods when normal science runs up against inherent limits, and my sense astonishment as open to correlation with Kuhn's "paradigm shift." Kuhn himself does not evince interest in religion and metaphysics, and yet there are religious and metaphysical hauntings there, even when the thinker believes his mind is on the positive, perhaps positivist side of their exorcism from his soul. (If there is a right side of being enlightened, presumably there is also a dark side, which the light casts into the shade.)

It might be of relevance to note that I explicitly resort in *Being and the Between* to discussing Kuhn as helpful in making sense of the interplay of univocity and equivocity, as well as of the transition from equivocity to a more dialectical orientation to being.[13] Some of the relevant thoughts there were the following. I noted how some critics understood Kuhn's claim of incommensurability as a derogation from the scientific task into an equivocity that itself was ambiguous. Kuhn compared incommensurability to a *Gestalt-*switch; "it must occur all at once (though not necessarily in an instant) or

13. Desmond, *Being and the Between*, 123, 125–26.

not at all."[14] One now sees a rabbit in the *Gestalt,* one now sees a duck, and there is no intervening Gestalt that allows one to pass from one to the other in a mediated way. The movement from one to the other is abrupt—it seems unmediated. As with the equivocal sense, the stress falls on difference to the exclusion of mediation. Hence, Kuhn has been attacked as a relativist, sometimes with a fierceness that calls to mind an earlier claim of incommensurability, namely the Pythagorean discovery of irrational numbers. Granted Kuhn was not killed, as supposedly Hippasus was for revealing the irrationals.

Relevant to the current discussion is the fact that Kuhn has been attacked for using the language of *religious conversion* to describe the shift from one paradigm to another.[15] This is a touchy thing. Unlike the Pythagoreans, the standard picture of religion by the scientistically tempted is that religion is *the* domain of irresolvable equivocity. The claim of incommensurability seems to deny any univocal foundations to the scientific enterprise, and indeed to undermine the confident claim that scientific reason is somehow exempt from all that murk. Kuhn does specify other considerations that enter a decision, such as accuracy, scope, simplicity, coherence and consistency, fruitfulness. None of these "standards" seem to be unequivocal criteria. Kuhn calls them norms of *value* rather than univocal criteria. As such they have a constitutive ambiguity or perhaps open-endedness. The scientific enterprise seems to be infected with the equivocity from which it seeks immunity. The "standards" supplementing the more recognized univocities carry their own potential for equivocity. The "standards" said to enable us to transcend "wooliness" are themselves "woolly."

Though Kuhn did come to qualify his views, one notes the equivocity in the fact that some attacks on him have been extreme, as if the "problematization" of univocal foundations leads to an abyss of nonsense, with science itself being dethroned. True, his way of talking of "revolutionary" science has been adopted by some with the aim of glorifying a kind foundationless equivocity in other areas of human life. I think that the philosophical issue is less the loss of univocal foundations or the apotheosis of equivocity, but the search for grounds of intelligibility in a manner that preserves the truth of univocity and equivocity, while transcending both. We are pointed in a more dialectical and metaxological direction. That we need more than the univocal sense does not mean we do not need that sense. This means an affirmation of curiosity as a modality of wonder, as well as a reaffirmation

14. Kuhn, *The Structure of Scientific Revolutions,* 150.
15. Kuhn, *The Structure of Scientific Revolutions,* 148.

of the other modalities of astonishment and perplexity that we neglect at our peril.

What of the importance to granting a difference between scientism and the practices of the sciences? Andrew calls on his own experience of professional scientists and finds that they are not always scientistic. When they are, this is bound up with the scientist's cultural, educational, and social setting in middle-class society. This is a point worth stressing and pursuing, particularly with respect to what Michael Hanby and David C. Schindler say, both of whom, and Michael somewhat more strongly than David, find it problematic how to make at all the separation between science and scientism. I will say here that Andrew Davison's stress on *practices* seems to me right: we need to attend to what scientists *do*, and not only to what they say they do, or say about what they do. Indeed, in what they do they may be, so to say, *incognito* witnesses for originating wonder, even if their reflective articulation of its meaning may come out in scientistic or quasi-scientistic form. They know not what they do. It is true that the obverse holds also: namely, the language of (self-)understanding and (self-)justification may speak the truer and more ideal language of originating wonder, but the reality of the scientistic deed will mask the insinuations of secret will to power in its potentially poisoning power. I mean that noble science will be the cover under which ignoble deeds are allowed to pass without demur. I think the concept of the counterfeit double again here works in many registers. I will connect this below with the parable of the wheat and the darnel.

Kuhn's stress on history brings more to the fore the social reality of modern science as itself historically instituted and impossible to dis-embed entirely from the historical ethos in which it participates. There will be permeability between the ethos of the social in a more encompassing sense and the practices of science shaped and specified by its own socially mediated institutions. This is an issue at play in Simone Kotva's discussion also. If the ethos more generally is too defined by the dominion of serviceable disposability, the scientistic blandishment will be less easy to resist specifically. This is relevant to Andrew's suggestion that what scientists say about their practices is often shaped more by their place in polite, educated, middle-class society than it is by their work as scientists. Yes, in what they do they are not necessarily scientistic; the interpretation of what they do under the dominion of the *Zeitgeist* of serviceable disposability tends to offer words for self-interpretation that are more scientistic in character. There is often a disjunction between the practice of scientists and what they reflectively say they are doing.

Something that has struck me in connection with Kuhn is the almost *holy* air surrounding the words "revolution," "revolutionary." Perhaps this has been so since the French Revolution, and other unholy revolutions. When a word becomes a cliché in the advertiser's dictionary of triggers, as "revolutionary" has become ("revolutionary" new washing powder, . . . revolutionary yada yada yada), then a lexical virus has already infected the *Zeitgeist* and one can be a carrier of it unbeknownst to oneself. If it is a marker of importance if something is labelled "revolutionary," it is important to ask if and how this then did affect the reception of Kuhn's ideas, outside the more specialized domain of the history of science. My surmise is that Kuhn's "fame" in part is due to the revolutionary resonance of the times and our modern times—and the sometimes spoken, sometimes unspoken view that we can remake all things anew. I do not think we are yet that kind of god. The serpent of the scientistic tempter preys on our willingness to be seduced by everything "revolutionary."

The disjunction of practice and reflective account never entirely keeps at bay the temptation to scientism in a more strict sense. The dream of being a determining god is not absent in the laboratory, as it is not absent in the laboratory of revolution. When the scientist brings their more characteristic forms of determining thought to bear on being, the determinability of things comes back to haunt the practice itself. Practicing scientists are beneficiaries of what sometimes their practice confirms but which it sometimes denies. Scientism is a philosophically inspired interpretation that tilts in the latter and not the former direction. It finds it hard to endure the equivocity and plurivocity of being. This is why Andrew Davison is right to sense a gap between Kuhn and myself with regard to the strength of the "realist" stress I would advocate. The "naïve" realist is often criticized, but one might criticize the naïveté rather than the realism, even though sometimes idealism claims victory in the disposal of realism. But our desire to be true to being is native to our own being, and if this nativity is the meaning of naïveté, then there is no escape from it, except into lunacy, but even then lunacy is only truly lunacy because of our nativity to our being in the truth by our being called to be truthful.

If I can again refer to *Being and the Between*, I make some remarks relevant to Andrew's questions about realism. In the end, mediating reason is not to be abandoned, even as mindfulness undergoes a complexification beyond the first univocity and the return of the equivocal. The search for univocal *mathēsis*[16] may contribute to more comprehensive modes of determinate thinking. Renewed fidelity to concrete being makes demands on our minding that it activate its mediating powers in more dialectical and metaxological manners, and in ways both scientific and extra-scientific.

16. Meaning: learning, science, knowledge (esp. mathematical).

Certainly any scientistic self-image of science as a triumphant univocity that puts all other modes of minding in their subordinate place has to be given up. A new community of mind is asked between science and other modes of mindfulness, a community that itself is respectful and supportive of the difference and justified ways of these other modes of mindfulness. Scientific mind becomes a member of a complex community of ways of being mindful. It ceases to be the imperialistic master, as it has seen itself in scientistic interpretations. Andrew is not wrong. Whether or to what degree Kuhn accepted or would accept friendly amendments in a more "realistic" direction is another question, but one could argue that if one were to defend some distinction between science and scientism, the philosophical interpretation comes back into consideration, as to whether we offer a more "realistic" view of its cognitive claims, or a less. Some metaxological realism is unavoidable, in my view, and this by no means obviates rejection of insights that emerge in orientations that have a more idealistic, or pragmatist, or phenomenological character. Much depends on how we conceive of being true, and I have attempted to give a metaxologically pluralistic view of being true in *Being and the Between*.[17]

To reiterate: Kuhn's insights are suggestive for the impossibility of giving an entirely univocal account of the nature of scientific knowing and development, and I am in Andrew's debt for elaborating on significant convergences and divergences with my work. I am not sure I would refer to perplexity as having a rather "dour role" to play. I do think we can be tortured by perplexities, especially those we find hard to answer and yet impossible to abandon. In perplexity one might be torn as to what might constitute an appropriate answer or response. I think of tragic situations as precipitating perplexity; I think of ethical, metaphysical, and religious perplexities. One realizes that one is elsewhere than in the commensurable space of scientific/scientistic problems or puzzles. I think Kant had a sense of this with regard to metaphysics. This reveals the non-scientistic side of Kant. One wonders if the culture of any scientism is able genuinely to acknowledge the tragic, or perplexities turning on evil, its devastation, its forgiveness, its redemption. I take this to be a serious issue, requiring not only non-scientific vision but entirely at odds with the thoughtless faith of scientism in its own cognitive powers and their promise of revolutionary perfection. Revolutionary perfectionism, like scientism, flees the tragic dimensions of the human condition and betrays, while counterfeiting, our trans-political longing for religious redemption.

17. See Desmond, *Being and the Between*, 128–30; also ch. 12 on "Being True."

VII

I am grateful for **Richard Colledge's "Scientism as the Dearth of the Nothing"** for his very attentive, very thoughtful, and sympathetic treatment of central issues in my reflections, which I very much appreciate. He draws our attention to sometimes similar, sometimes slightly dissimilar, themes in Heidegger, and perhaps with the gentle nudging towards a more sympathetic appreciation of Heidegger, especially his later work. One need not see our diverse sympathies as antipathetic, and yet I do confess at times to a not quite dispelled antipathy, to a need to "step back" from something not entirely on the level in Heidegger. This has grown with me.

The contrast of earlier and later Heidegger has occupied many commentators. To me there are many Heideggers, not simply pre-*Kehre* and post-*Kehre*. Heidegger, like *to on*[18] in Aristotle, and indeed as I would say *to meon*, the nothing, is *legetai pollachōs*, said in many senses. In connection with his philosophical relation to the religious and the theological I counted at least seven Heideggers. I do not have space to enumerate them here.[19]

I agree with Richard's handling of the *Stimmung* of especially the later Heidegger in many ways, and some of the things I would want to say resonate with what is admirable in the later Heidegger. Poets and philosophers can each be consecrated priests of astonishment. And one of the great gifts of Heidegger, early and late, is his power to raise questions, old ones anew, and new ones in ways with connections to and disconnections from the old, putting us on a threshold when the promise of something not yet quite heard is hinted. What persists in the best of his philosophizing is the phoenix of original astonishment. At best it is this astonishment at the "that it is at all" of being that trumps anxiety before the "not-being-any-more" that is our mortal destiny. I say the *practice* of his philosophizing, though certainly the early Heidegger does give the lead role to *Angst* on the (me)ontological stage.

I think of the attractiveness of the later Heidegger relative to *Gelassenheit* and his not-to-be-avoided reflections on technology. This is very relevant to the theme of scientific curiosity and scientism. There is an overlap between what Heidegger calls "standing reserve" and what I treat of in terms of "serviceable disposability." I take the earlier Heidegger as recuperating metaphysics in reawakening the question of being, but the later Heidegger is more in the business of overcoming of metaphysics as ontotheology and

18. Meaning: being.

19. Desmond, *Godsends*, 5–6. I think too of the theologians in relation to the earlier Heidegger in *Being and Time*, a book full of de-theologized Augustine and Kierkegaard: they inhale the lingering perfumes of the banished god.

as the historical enabler of the reign of cybernetics. One might use my own way of speaking of being between system and poetics. The Heidegger of *Being and Time* is in the business of generating something like a version of transcendental-ontological *Wissenschaft*, while the later Heidegger's dialogue with the poet, especially Hölderlin, turns to the conversation of thought with the poetic. I see this last as less unprecedented than some of Heidegger's admirers claim, since it is in the same line as the interplay of art, religion, and philosophy we find in the acme of classical German philosophy, with Hegel and Schelling. That Hölderlin was the significant poetic third to Hegel and Schelling in their own intellectual ventures makes something of the point.

I also see our need for metaphysics differently: metaxological. I am very aware that Heidegger does refer us to *das Zwischen*, but our under-standings of this are not coincident, even if there are overlaps. There is no systematic exploration of the *metaxu* in all the dimensions I have explored and articulated. What Heidegger calls the "step back" (*der Schritt zurück*) can be endorsed, but taken differently. After all, the "step back" does con-cern what makes metaphysics itself possible at all. This is important since Kant's transcendental turn, which also asks about enabling (or disabling) pre-conditions, though it is a prolegomena to any future metaphysics rather than the overcoming of metaphysics. Heidegger is right in moving from transcendental conditions to ontological enabling, in raising again the ques-tion of being in regard to ontological and not just epistemological enabling. I would endorse this move.

Wonder as original astonishment is also at the origin of metaphysics, but is there possible a "step back" from that astonishment? Perhaps only in the sense of us becoming more mindful of what enables the astonishment at all. Step back, yes, but in a different sense to Heidegger. I am surprised that Heidegger did not see fully the promise in the adage of Aristotle, surely well known to him from early on: "Being is said in many senses." The later Hei-degger risks essentializing the tradition of philosophy as all but a univocal totality in terms of ontotheology. While he is not entirely wrong in drawing attention to ontotheology, especially with respect to Spinoza and Hegel and more generally to the rationalistic tradition in metaphysics, to speak of the essence risks a trans-epochal univocalization and an epochal totalization. I fear that Heidegger's totalizing of "the tradition," while reminding us of the forgetfulness of being, risks producing a *second forgetfulness of being*. And this, not only in the way of his admirers who remember Heidegger's texts more than being, but also in the effect of the venture of overcoming metaphysics which induces its own forgetfulness of being.

I would put the stress on the plurivocity of practices of metaphysics, metaxologically rethinking Aristotle's "being is said in many senses" in terms of the fourfold sense of being. This is a needed philosophical task whether one claims to return to the pre-metaphysical, or to practice metaphysics along the lines of a more traditional discipline, or claims to succeed or exceed that practice in overcoming metaphysics. Be it one or the other, all will be shaped or guided by some one or more of these fundamental senses of being. This is true of the post-metaphysicians—they remain metaphysicians.

Heidegger's efforts to recall a sense of truth more primordial than propositional truth is an important consideration, and I think is relevant to the issues to be raised about scientism. Scientism can be looked at as a fixation on determinate or determinable truth in a manner that occludes a prior determining that enables determinate truth to be true at all. When I speak of porosity to the true, Heidegger's discussion of truth as *homoiosis* and as *alētheia* seems especially appropriate with respect to the determinacy of univocal truth and the indeterminacy of equivocal truth/untruth. Heidegger is not wrong to call attention to a sense of being true more primordial than propositional truth; we must presuppose our already being in a porosity to the true for us to be able to determine this way or that the truth-worthiness of this determinate proposition or that. I would say that with regard to the plurivocity of being true the matter is not just the doublet of determinate truth and an otherwise indeterminate hiding/unhiding that is said to be more primordial than propositional truth.

Perhaps in tune with Richard's theme of "the dearth of the nothing," Heidegger's earlier statements about *alētheia* explicitly stress the *privative* nature of the unconcealing. *To alēthes* is the unhidden—*das Un-verborgene*. *Alētheia* is the privation of hiddenness—*Unverborgenheit*. Heidegger even likens truth to being always a robbery (*immer ein Raub*) when speaking of the uncovering of truth in *Being and Time* (§29). In the original German Heidegger stresses the word "*Raub*." What kind of theft is this? Is not robbery taking hold of something over which one has no rightful claim. I am not familiar with any commentator dwelling on this. What would an ontology of robbery be, what a theology of theft? (I am put in mind of the theft of Prometheus; Prometheus is invoked in Heidegger's Rectoral Address, the only place in his work Prometheus is invoked, according to Hannah Arendt.) The primal porosity of our being true is not a matter of an oscillation between determinate and indeterminate. A privative unconcealing is not quite true enough to the overdeterminacy of the mystery as giving the porosity, and enabling our *passio essendi* as itself a *conatus*[20] for the true. I venture that this *conatus* shows a kind of ontological "connaturality" between our being true and the true as other to us, and always as overdeterminate (not just indeterminate) in excess of our determination and self-determination. I would call this agapeic surplus rather than negation of a hiddenness and privative unhiding. Its gift is not to be "robbed."

In Heidegger, what seems like the equiprimordiality of hiding and revealing seems finally to tilt asymmetrically towards a more primal hiddenness. This might be true, in one sense, (to speak theologically) for the hyperbolic God who dwells in light inaccessible, but not quite so, for nature naturing, or nature natured; though in another sense this hyperbolic God is nothing but self-revealing, even granting the asymmetrical transcendence of the divine. This God is not the Heideggerian origin.

The "dearth of the nothing" can be connected with scientism. In his inaugural lecture, "What is metaphysics?" one can appreciate Heidegger as highlighting the nothing as a provocation to those, especially scientists, who zone in on determinability, and a chastening to any totalizing ambition in this. The reminder is that the sense of being as a whole might come to be encountered as a question against the meontological background of the nothing. Heidegger is getting at a different sense of nothing, other than determinate negation or self-relating negativity in Hegel's sense. I think here there are the different senses of the nothing, broadly corresponding to the overdeterminate, the indeterminate, the determinate, and the self-determinate. Heidegger points beyond the last two but, I think, equivocally mixes the first and the second. I have tried to outline these different senses of nothing elsewhere, in a reflection entitled "God Being Nothing More."[21] Augustine asked himself: *deum et animam scire cupio. nihilne plus?*[22] and answered *nihil omnino.*[23] I want to offer an ontotheology of the Plus in relation to Nihil—ontotheology perhaps, but not at all in Heidegger's sense. Just as being is, so nothing is said in many senses. There is a sense of nothing correlative with the overdeterminacy of being, a *nihil per excessum*[24] (as Eriugena might call it); a sense of nothing correlative with indeterminacy, a

---

21. In a response to Ray Hart's book, *God Being Nothing.* The volume in which "God Being Nothing More" will appear is edited by Alina Feld and scheduled for release with Edinburgh University Press in 2023.

22. Meaning: "I want to know God and the soul. Nothing more?"

23. Meaning: "Nothing at all."

24. Meaning: "Nothing through excess."

nihil per privatum,[25] so to say; a sense correlative with determination (*negatio est determinatio*:[26] Spinoza yes, and perhaps this sense is the most easily recognized and the most recognized); and a nothing correlative with self-determination (Hegel's self-relating negativity would be a classic instance).

There is an interesting doubleness in speaking of the dearth of the nothing. In Richard's rich exploration it is something of the deficiency of scientism to be true to being, in its evasion of the nothing, that is the dearth. There is a philosophical dearth in a different sense, when one might speak of a kind of poverty of thought, just in its always failing in coming to terms with the nothing, a failing that is strangely more of a genuine success than all scientistic successes. Poets fail scientistically, but in this other dearth of the nothing they may succeed in being truer, say, to the tragic dimension of human existence. I have myself spoken of a new "poverty of philosophy,"[27] beyond our claims systematically to consummate rationality. Richard importantly recounts Heidegger speaking of "the poverty of reflection." The tone has nothing to do with theft but rather access to "the promise of a wealth whose treasures glow in the resplendence of . . . the inexhaustibleness of that which is worthy of questioning."[28] Richard rightly reminds us that here thinking has an essential kinship with thanking: "*Denken/ Danken.*"[29] It is to be acknowledged that "by the mid-1940s Heidegger is writing about 'the transition out of willing into releasement [*Gelassenheit*],' with what he now calls 'thinking' as 'releasement to' the 'open-region' of being."[30]

I still think Heidegger does not address the question: How get from *Polemos*[31] to *Gelassenheit*, from theft to gift, from taking to being graced? While one can grant the second, one wonders how it is possible at all, or whether perhaps the second reveals really something more primal, signifying that the matter should be inverted: not from theft to grace, but from grace to theft; signifying a father more primordial than pater Polemos, a father who, if so granted, asks for an entire *metanoia*[32] in our thinking, indeed also in the practices of metaphysics. I see metaxology as a metanoetics in that regard in which the agapeic origin comes for consideration not simply

25. Meaning: Nothing through privation.

26. Meaning: Negation is determination.

27. See Desmond, *Is There a Sabbath for Thought?*, ch. 3.

28. Heidegger, *The Question Concerning Technology and Other Essays*, 181.

29. Heidegger, *Pathmarks*, 236.

30. Heidegger, *Country Path Conversations*, 70, 80.

31. I am referring to the statement of Heraclitus: Polemos is the father of all things. Polemos has a kind of mythic-metaphysical significance for him.

32. Meaning: change of mind and thence of action (repentance).

as an equivocally indeterminate, perhaps overdeterminate origin but as the ultimate overdeterminate enabler of the porosity of being, within which the determinately true and intelligible, and the relatively self-determining come to be.

There is the dark middle of the 1930s when Heidegger speaks of the need of an enemy, and the need to create an enemy, if one does not have one. I see no thanking here. How get from enmity to gratitude? Was the darnel already in the wheat in the earlier writings, and did the darnel grow into an evil Nazi bloom in that phase of its unfolding? And though wheat begins to flourish again, is this in a philosophical purgation, never honestly acknowledged and confessed?

Who would deny the genuine pathos in the later Heidegger but is it enough? I find it hard to put out of mind a strange, perhaps inverted image of Comte relative to the history of being as onto-theologically determined: from theology to ontology to cybernetics. Heidegger's sensitivity for the equivocity of being as both concealing and revealing, and indeed concealing in revealing, even to the concealing of concealing in revealing, perhaps does serve him to step back, step away from scientism in the form of positivistic cybernetics. One fears that this last reveals the shadow of a kind of global projection of scientism. It is a projection to the whole, and if it is a totality, there is something counterfeit about this totality. Even here there is an equivocal intertwining of concealing and revealing, since the claim of scientistic enlightenment hides its own darkness, when it comes to the truer sense of the meaning of being. Perhaps this goes some way to explain Heidegger's *apocalyptic* sense, all talk about *Gelassenheit* notwithstanding. It is as if he appreciated deeply the need of a breakthrough beyond counterfeit forms of universality.

Perhaps there is urgency earlier to this that made him permeable to the pagan apocalyptics of Nazism. Perhaps later there is a different urgency of ultimacy of something coming through via the totalizing world of scientistic cybernetics. If the equivocal between is the first word and the last word, then the feeling of being lost or deceived or destined to erring can take hold. I find this understandable in that any beyond of such an equivocal totality has been closed off by the global projection of the cybernetic frame. The impasse is redoubled, heightened if all being is only equivocal. But has the "beyond" of the equivocal "universal" already been closed off at the outset, perhaps by the postulatory finitude of a self-enclosed immanence? The philosopher, like all of us, is in a hole, but he can also dig himself in a deeper hole. In the equivocal, counterfeit whole we are (all) in the belly of the beast. Globalizing scientism silences the very words that would release us to a truer and more intimate universal . . . and to a truer dwelling in the between.

Heidegger himself is on the right track when he says only a god can save us, but more than one god has failed: not just the god of communism, not only the psychopathic god of Nazism, but *der letzte Gott*[33] of burlesque prophecy. And it seems he himself has committed to silence about the name of Christ, and so a God that has saved is not named.

VIII

Jeffrey Bishop's "The Determinations of Medicine and the Too-Muchness of Being" shows us how precise attention and personal discernment need not be at odds. The reflection itself shows us how meticulousness and finesse are not mutually exclusive. The essay offers us haunting stories. It offers a marvellous meditation, hard to better, in which we find variations on themes in a musical sense: the thought moves up and down, and down and up, in and out, a song in the healing porosity between the doctor and the patient. It is true to life in confronting the failure of life, the frailty of our mortal condition. I see it as a piece companioning to my own thought in which I saw incarnated things I have tried to say, perhaps too abstractly. There is sustained attention to the intimacy of our being, to what I call the idiocy of being, in particular to idiot selving, everything about our singular being that cannot be entirely contained in the abstract generalities of the space of public communication.

The intimacy of incarnation is almost impossible to communicate but not impossible. It is much easier to dwell with more familiar determinacies. Jeffrey manages an incarnating reflection, giving flesh to the porosity of being, even granting that the flesh is itself the incarnation of the porosity. It pays attention to surfaces without being superficial: what surfaces in more intimate communication with others, especially in face of the other. The look of the other is most important. The *eidos* is the look of things, the Greeks thought, but I will leave that aside now. Humans are concerned with how we look, how others look on us, look at us. There is something metaxological about the look: it can be both subjective and objective, and yet also trans-subjective and trans-objective: the look on the face can communicate the finesse of the heart. Something surfaces in the face but it is not superficial, more the overdeterminacy in the frailty of the being as other and its astonishing being there. Overdeterminacy of being passes between human beings in the transience of the living look.

When Jeffrey Bishop speaks of science dealing with the general, and in search of a covering theory for multiple instances, while medicine zones in

33. Meaning: the last god.

on the particular, I am put in mind of very Aristotelian reflections about the difference between theoretical and practical reason. Practical reason deals with what could be otherwise than it is; hence a finessed attentiveness to the nuances of particulars and occasions is required. At the same time, the achievement of the excellence of practical wisdom is not a feeling and a simply immediate intuition but is an achievement of character into which the habit of just such attentiveness enters, making it a credible and faithful source of sound judgment, even when there is no covering universal into which to slot the particular. I take Jeffrey Bishop to be telling us a tale of the defects of this slotting, learnt itself first from defect of attention on the threshold between the doctor and the patient. The point here made is also relevant to Jonathan Horton's reflections on the law and I will return to it.

There is a patience to the doctor in the attaining of such patience to the patient. The patience of the healer is no mere passivity, but a highest energization of mindfulness to the full reality of the person before one. There is no need to depreciate the theoretical, and given that Aristotle is called as our companion here, I would argue that as the ancients knew *theōria*, there is a beholding of a purer sort at stake here: a "beholding from" and not just a "beholding of." I mentioned this above in response to Steven Knepper. The "from" indicates the communication from the other to us, while the "of" primarily tends to be taken to mean the communication from us to the other. One is laid open, one lays oneself open, in being opened in patient mindfulness to the full being there of the other in his or her otherness.

I think the modern duality of theory and practice too constrictedly speaks of mere theory while often priding itself on its more practical, useful achievements. But if there is not porosity and the *passio essendi* in our *praxis* then we are just imposing ourselves on the other before us, and our *conatus cognoscendi*[34] overrides receiving. In fact, modern "theory" is primarily not "beholding from" but the projection of a constructed hypothesis, hence "practical" in a new univocalizing way that risks aversion from "beholding from." I take Jeffrey's reflection to occur in a space where we must cease to think and act in binary, indeed bifurcating terms between the *passio essendi* and the *conatus essendi*.

Let me connect the point with Jeffrey's reflection on the gaze. Though Foucault is mentioned, what came to my mind is Sartre's account of the look. With Sartre, in a merely apparent dialectic, there is a bifurcating also: my look objectifies the other, and in advance disarms the look of the other as threatening to objectify me. Rather than a dialectic of mutuality with promise of differently being together, there is here a see-sawing of competing

---

34. Meaning: endeavour to know.

wills to power. In this bifurcated see-sawing being with the other, I must kill by my look the threat of the look of the other, in advance of being killed by that look. If Sartre was right, and if looks could kill, we would all be mass killers, at least in potency. Levinas, in a way, simply reverses this, and the appeal that comes from the look of the other becomes the command: "thou shalt not kill." An inverse see-saw of a dualism tilts the ethical primacy to the other away from the existential-ontological primacy of Sartre's self as being-for-self. In neither is there an agapeic "beholding from" in the look of the other, and indeed now too in my look towards the other. Levinas is superior to Sartre, I think, in respect of a just ethics, but his account is haunted, as is Sartre's differently, by a variation of Hegel's master-slave dialectic.

Jeffrey Bishop is trying to recall us to the promise of "beholding from," and indeed its promise of agapeic being within the practice of medicine itself. The concept of "care" is transformed from a concern with outcomes and results into a care for the intimate being of the person before one, impossible to instrumentalize in terms of categorizable outcomes but just there in its entirely worthy being for itself. Yes, care in the more instrumental sense of proposing the medical means which will help or alleviate the "patient" may well be forthcoming. But this *other care* as being in the company of the suffering singular person, and being with their distress, being with them in distress, is care for the patient in which it is the carer who is patient to the patient. One person in the exchange does not exhibit *conatus essendi*, the other *passio essendi*. The *passio essendi* is on both sides of the metaxology of medical practice and reception, the metaxology of the flesh in suffering and healing, and indeed in dying. Jeffrey's meditation is on how he come to understand this metaxology in return to the porosity of his own being, by being there differently with the woman.

Again, to look at our eyes: There is a contrast of *studiositas* and *curiositas*[35] in some contributors, and it is worth recalling that we did not have to wait for Sartre's look to acknowledge issues at stake. I am thinking how the violence of the eye was always recognized by anyone who is attentive to the tender intimacy of the human being and its soul. The Christians spoke of *custodium oculis*.[36] There is a violence of the eye in scientist curiosity. The need of this *custodium* is with the aim of creating a sanctuary for the reverence of the soul in relation to other things and itself. Far from being a turn away from things and persons, it is rather a deeper, more interior care lest the tyranny of the eyes violate other being. It leads to the practice of looking with the more tender look of love. Those who love can hold the look of the

35. Meaning: "studiousness" and "curiosity."
36. Meaning: guarding the eyes.

other, can be held in the look of the other, longer and more truly than those for whom the look is a pugnacious, insolent, warlike "eye-balling." We see the latter in the sadistic foreplay of the professional pugilists, trying with their weaponed eyes to steal an advance submission of the other before even the actual fight begins. The lover does not eyeball the beloved but looks with vulnerability and appeal and a kind of shy adoration. It is the rapist who is the eye-baller, and as there is an adultery of the soul, there is also a raping too. I recall Galileo's praise for a "rape of the senses" in the new mathematical science as he understood this.

It is amazing how sometimes a look is enough: a whole world, the whole person can be communicated in a look, understood by one from the look. Jeffrey beautifully reminds us of this. When I was young in the 1960s, there was a love song hit by a group called the Hollies in which they say that all it took to fall in love was "Just One Look." Not quite marked by the deep pathos of Jeffries's account, I had the following experience with an oculist when my eyesight was decaying due to glaucoma: after tests with multiple marvellous machines, my doctor looked at the charts and the computer pictures, and pronounced that all the numbers were good. I said: the objective numbers may be good, but subjectively my sight does not feel as good as the numbers look, and I am not seeing well. The doctor was surprised, even astonished at the way I talked. He gave me an odd look. He was struck by something; and afterwards in different visits, he looked at me somewhat differently.

I mention how the intimacy of our fleshed being is not unconnected to the theme of the counterfeit double. The pervasive claim to be concerned with the victim sometimes reveals counterfeit doubles of agapeic service. Counterfeit compassion for life as disposable: abortion. No child of wonder born: life's disposability not only in the killing in the womb but in the harvesting of body parts and foetal tissue—all justified as the compassionate betterment of our lot. Consider the dimensions on which human flesh is disposed on the industrial scale of modern abortions, a scale made possible by the efficiencies of modern science and technology. There are counterfeit doubles that trade in the systematic desecrations of the flesh. Counterfeit compassion for the dying: euthanasia. No saving sacrifice of the Child of Wonder.

"The body may have its reasons for pain, of which reason knows nothing." Jeffrey Bishop gives us good reasons for remembering these reasons. Pascal incarnated: The heart has reasons of which the scientist knows nothing. He speaks of how he came to know that the burden of his gaze, itself weightless, increased her burden, and brought to the surface awareness of the weight of her situation, there all along but not surfacing. There was a nameless passage back and forth between doctor and patient. He very rightly draws our attention to the weight of the gaze, something that should be weightless, and yet it weighed on her. Is the Augustinian thought helpful here: my love is my weight, *pondus meus amor meum*? The loveless gaze is also a burden but not the loving gaze that ties us together in a bond that is not bondage.

I think of Monica with her tears. The tears of the woman who suffered spoke. Speaking with tears—wording, watering her suffering being with tears. When tears speak, they plead. They ask of the doctor questions that impel the healer to question *themselves*. We are in the neighborhood of compassionate mindfulness. We speak of *lacrimae rerum*, the tears of things. They cry to us, and when we grow frail, they cry for us. Persons too are things that cry. And while the beatitudes tell us that those who weep are blessed, there is weeping that is the vestibule rather than the sanctuary of blessing. Jeffrey Bishop is blessed in this vestibule here. The wording tears of the sufferer open a tear in the healer, a tear that perforates the doctor closed in on himself, making him newly open in suffering the suffering of the other. This is medical metaxology as a *compassion essendi*. There is a healing intermediation: the senses interpenetrate—a touch is a tear, a comfort, a look is flesh not stone, a sigh is breath loving and grieving and longing. . . . All touched on, touched by tender finesse, in this essay. Its words are worthy to be read in their own eloquence and not just paraphrased.

IX

Jonathan Horton's "Attending to Infinitude: Law as in-between the Overdeterminate and Practical Judgment" puts me very much in mind of some of the things I have tried to say in *Ethics and the Between* about the potencies of the ethical, particularly what I call the *dianoetic* potency. This dianoetic potency reflects the need to seek intelligible, law-like regularity amidst the ambiguities of our lives. It also puts me in mind of things I say about the eudaimonistic potency, in so far as this tries to hold together the aesthetic and the dianoetic, as well as fostering the sense of practical wisdom, *phronēsis*, as significantly important as staying true to the nuance of the particular without betrayal of the more-than-particular at the level of binding and applicable law. The true judge "betweens," taken in the verbal sense of intermediating and coming to a just determination between the general and the particular.

Our lives are full of equivocity, yet as we become discerning, we discover constancies and regularities. Ethical and legal systems reflect our respect for these constancies and regularities. Contrary to a certain postmodern tendency, constancies and regularities need not be reductive of diversity. There is the temptation, to be sure, to the superimposition of orders on the surplus of the overdeterminate, now disciplined in its tendency to waywardness. And the law can sometimes become tyrannical in dictating to the ambiguity of living situations, and yet law is needed still to rightly order that ambiguity. The law can kill through false fixations on rules to the detriment of justice. Not doing justice, such spurious dictations can be counterfeit doubles of dianoetic constancies that shape our ethical being and enable lawful life.

Jonathan Horton's discussion is a very striking and illuminating application to the law of the interplay between the overdeterminate/indeterminate and the determinate, not exclusive of the presence of the self-determinate. We are called to attend to the requirement of a kind of finessed betweening intermediating the more general and the more particular, a betweening that is not either one or the other, and yet can be embodied in a just judge who intermediates between the one and the other. The essay is itself a witness in its own performance to a kind of reflective and responsive betweening: between some of the seemingly general thoughts I offer and particularities of the law and its practise(s), extending the reflections I offer in a direction that illuminates legal situations in the round; between the high horizon of ideal justice and the often un-wonderful falling short of justice in particular actions or persons or groups or circumstances.

As with my remarks about the healer in Jeffry Bishop's essay on medicine, the importance of someone like an Aristotelian *phronimos*, a person of practical wisdom, is at the center of the betweening relation of overdeterminacy and legal determination. Such a person does not drop from heaven ready-made but is the educated outcome of a life of mature experience, as well as in this case, a pedagogy in a legal system, itself refined through a historical unfolding in which finessed judgment of the particular is required along with knowing of the generalities of the law itself. The legal system in question is the English common law rather than the continental and it would be an interesting question further to pursue as to whether the latter shows more of the *esprit de géométrie* than the *esprit de finesse*. The latter distinction belongs to a great Frenchman, but it mirrors the Greek reverence for *phronēsis*, as well as the English respect for common(sense) prudence.

It is interesting to think that an (Anglo-)Irishman, Edmund Burke, brought to the fore the worry about the danger of the abstract "metaphysicians" and their rational projects to create, as if from nothing except human power, the new law of a revolutionary politics. At work in Burke's worry was a realization of the necessity of such prudential finesse, and the danger to the stable order of justice of geometrical dictation from above of fixed, univocal regularity. Paradoxically, a wise tolerance of the equivocal allows a supple dianoetic constancy in the formulation and application of law. By contrast, an insistence on univocity of the more geometrical sort can give rise to a dissolving, if not self-destroying equivocity in the legal and political system.

Mindfulness that "does justice" is attentive to the nuances of being in the between, and informed by a "taste," a "*savor*" for the "whole." Savoring mindfulness leads to more particular discernments in the plurivocity that find out ways through the chiaroscuro of the ethos. The needful discernment is neither a fixation on rigid principles nor a lax yielding to every passing impulse. It is principled and yet ever vigilant to the nuance of situation. One is talking of the genuine judge as a *living constancy* in the midst of the passing. Needed is an intermediation with the passing which is mindful of the constant, and how it gives direction in the midst. This does not mean the automatic application of univocal principles to this particular situation. More often the savoring mindfulness has passed into the second nature of habit, and is enacted seemingly without thought. It is intermediating vigilance become as like immediate instinct. The person of practical wisdom is simply the embodied discernment of what right now is the appropriate response or course of action. Being this embodiment telescopes a long history of intermediation with others, familial, fraternal, societal, and so on.

This does not obviate an *improvisatory wisdom*—the ability to face the new, and with new eyes to discern the proper that yet has no precedent. To be intermediating in the between is to meet the surprise of the unexpected, as well as the confirmation of the well-tried. Since this middle mindfulness only exists as concretely incarnated in this person, it is a living mindfulness that has its being simply as responsiveness to what is called for in the situation. It has an *original side* that probes the unknown, as well as a *conserving side* that confirms and guards what is worthy in what has already been. Such practical wisdom is what it means to be between ethically, and here by extension legally, in the ambiguity of the ethos, both in its general encompassing character, and the specificity of this particular situation. This middle mindfulness is not a merely domesticated prudence, calculating its own best advantage, and otherwise secure in a life of contented mediocrity.

Jonathan's fascinating essay does not scant on reminding us of the *awesomeness* of the law; as well as the plurality of considerations that must come into play if we are truly doing justice or bringing justice to bear on a situation. There is something of ritual in the legal judgment, even "a moment of sacramental significance," when we ask for a legal determination that brings justice to bear on a contested case. There is an intermediation between the overdeterminate and a pressing determinacy. This is more than forensic univocity, for even when a case is very univocally clear, the application of the law, the making of a determination, requires an intermediation that is more metaxological than univocal.

I am struck by Jonathan's reminder of how in some streams of influence, such as with Bentham's claim that legislation is a *science*, the pole of determinacy is so foreground that the complex intermediations are recessed, if not forgotten or denied. If we speak with Bentham and reduce the matter to a scientific approach we make the law vulnerable to instrumentalization. In the language I use, it becomes captive to the dominion of serviceable disposability, and the call to do justice is contracted. That the insinuation of that orientation to the law is to be found in Bentham and his inheritors tells us a lot about the permeation of the modern sense of reason by the spirit of univocal calculation rather than the finesse of practical reason in a more Aristotelian sense. Jonathan's reflections are reminders that much is legally at stake in the mutation of the meaning of reason in modernity in a more instrumental direction.

While at best laws are promulgated on the basis of discerned regularities, sometimes a system of rules can "lift off" (so to say) the life they are said to serve and assume a divorced life of its own. Law then becomes a self-enclosed system that does not release our discerning judgment but dictates to the human situation, as if from above. From above, the laws become imperial dianoetic units that charge the ethos and ourselves with conformity. We might think we are in the company of the higher and perhaps supreme good but something is out of kilter. The essay strongly affirms a different sense of "aboveness" to the law. It is not a matter of human claims to being superior, or sovereign measure over all. "Aboveness" means that we as judges are ourselves judged by a measure higher than ourselves.

Nor, given the need of betweening, is this a matter of a dualistic transcendence. The more-than-determinate is in intimate intermediation with the determinate, and indeed is asked for by the immanent meaning of the law itself, which entails it be both applicable and be applied. But once again making a determination cannot be reduced to univocal determinacy alone. Burke comes to mind again in relation to the to him "metaphysical" abstractions driving the French Revolution. Time, custom, the wise habits of a people in its history are not to be erased in the swift guillotine of a political project. I am put in mind of Jonathan Swift, another member of the Anglo-Irish Ascendency whose ironical recounting of scientific, political, agricultural, and pedagogic projects reminds us what absurdity comes from the loss of common sense. (See below my response to Simon Oliver.) Contra Descartes, this becomes less widely distributed the more geometrical projects reconstruct the ethos of our lives.

"Aboveness" brings to mind the question of measure beyond our measure. I am thinking of this complication: we may claim to be the measure of everything other than ourselves, but are we the entire measure of ourselves? Do we stand under a measure superior even to our claim to measure all things? My worry about scientism is contained in this question: the will to totality in scientific form is of piece with the will to power in moral and political form and its trend to tyrannical social form. The figure of Burke haunts our response in this regard. I think his recoil from the French Revolution had to do with a recoil from the counterfeit claim of "aboveness" by powerful humanity, and the driving belief that social, political, legal, ethical, familial orders could themselves be our "projects." What we project will always cast before it the shadow of evil that lurks in the seed of darnel in human wills. This seed, sown in society, can grow into a shadow over the whole, a totalizing shadow that blots out the living light of true "aboveness." The celebrations of the liturgies of the law as a kind of civil sacrament makes one want to purge the intoxication of politics in the form of usurping absoluteness since the French Revolution. When "aboveness" becomes power over religion, it leads not only to the weakening or simple loss of the trans-political but to its being guillotined in the name of the supremacy of humanity. This is an inversion of what Jonathan brings to our attention, mentioned above, namely our encounter with ritual at the point of judgment and "a moment of sacramental significance."

There is a further thought here I can only mention, namely the family relations of scientisms and the religion of humanity. Comte comes to mind again, this time as a figure much admired by Mill, himself as a child groomed to utilitarian adulthood by his father, apostle of Bentham. True, Mill did have what amounts to a nervous breakdown in passing from youth to maturity. That he recovered and was healed with the aid of the Romantic poets is not insignificant. Poetry gave him rebirth and a second childhood, perhaps. But was the seed of the first fostering of the utilitarian grooming entirely fallow subsequently? Earlier education in making the child calculative in a utilitarian sense produces a changeling. Through poetry there strikes the need of the return of the child of wonder. Did the gene of the changeling ever die in descendent children in the inheritance of serviceable disposability?

X

Simon Oliver's "Life's Wonder" offers an inviting opening in calling to mind a reversal of modern and premodern: beginning in wonder, ending in wonders as curiosities, moving from love to control. The whole is a thoughtful, balanced, and filling in of this reversal, making it evident that from seeming innocent beginnings something unfolds not at all innocent. I find this adds importantly to my own sense of the contraction of wonder that can feed the counterfeit double of wise knowing we find in scientism. Simon Oliver importantly points to the loss of the proper object of science in the pursuit of control instead of truth, and he sees this at work from the outset of modern science. This is to be found even when there was lacking the technological apparatus to make effective this desire for control, signalling an immanent mutation of curiosity away from the inherent exigence of wonder towards the truth of being as given. There is a dearth of wonder and reverence, even as a contracted wonder is exploited towards ends feeding scientistic curiosity. He brings to mind appropriately the distinction of nature and art, the inanimate and the animated and he is undoubtedly correct in reminding us that when we deal with "life" the overdeterminacy of being must be brought into the picture. Having eaten of the tree of knowledge, we want to eat of the tree of life. The first fall generates "projects" that invite a second fall in the project of life mastered: redone in being undone, undone in being redone.

One of the thinkers he invokes is Hans Jonas, whose importance as a thinker about life is crucially valuable for the issue of scientism. Jonas, not irrelevantly, is a philosopher with significant work on Gnosticism, as well as on science, especially biological science. Far from falling for any scientistic temptation, he gives us resources for an ethics inclusive of the worthiness of life and, though he might not quite put it this way, the good of the "to be."

It is very helpful to be reminded of the "violent, impious, or sacrilegious mode of curiosity" that was of concern to the Stoics and Augustine. *Curiositas* establishes idols with a view to prurience and control; its opposite, *studiositas*, is fearfully reverent before its subject conceived in the light of wonder. Simon Oliver is very aware how earlier thinkers were wary of curiosity in its sometimes seductive equivocity. The Stoics and Augustine were attuned to some kinds of questioning as possibly the *impiety* of thinking. How attuned was Heidegger to this when he speaks of questioning as the piety of thinking? The same question can be put with hatred as well as love, with Pilate-like cynicism as well as expectant earnestness. *Studiositas* is to be the more reverent deployment seeking to know, a point acknowledged too by Aquinas (David C. Schindler also reminds us of this).

I find it illuminating that he offers us evidence from the very foundations of the Royal Society. Perhaps I might add that, paradoxically the
irreverent spirit of the comic can be a family relation to the reverent spirit
of *studiositas*. I am thinking of Jonathan Swift and the Laputan travels of
Gulliver. I offer an image or two from Swift to illustrate the point, and also
to see that "projects" too can be endeavours of absurdity.

Consider Swift's wonderful satire of the project of *abolishing words*.
If words represent things, as some philosophers say, why not "shorten discourse by cutting polysyllables into one, and leaving out verbs and parti-
ciples"? Why not abolish the words altogether and just communicate with
things? Would it not save our lungs from corrosion and our lives no longer
shortened by having to speak words. We can converse by means of things,
carried around in bundles or if we are rich by our servants. We can speak
with someone by showing the other the things and reciprocally be shown
the things—wordlessly. The great advantage would be a universal language
since things are generally of the same kind throughout the civilized nations.
Swift adds: the women spoil it all, as do the ordinary people, by threatening
rebellion if they were not allowed to speak with their tongues in the old way
of their forefathers. So they prove themselves each, the narrator laments,
"constant irreconcilable enemies to science."

Swift's laughter is a great antidote to scientistic inclinations, mocking
the regnant project in modernity to univocalize everything, be it an educational system, a political or an economic system, a system of philosophy. He
shows the stupidity of univocity, just in its intellectual homogenizing, and
its insistence: one thing, one meaning. The resulting "conversations" show
that instead of intellectual transparency, without finesse univocity breeds
stultification. Swift anticipates the later Wittgenstein, a foe of scientism, in
some ways more effectively, with a few brilliant comic images.

Some of the projects mocked have contemporary resonance. I mention a project to generate sentences—anticipations of something like a
word-processor. The machine generates lots of broken sentences, though
these have yet to be made meaningful and intelligible. The projectors are
working on it. It reminds me of what was called a "postmodern generator"
some years ago, which could generate articles or essays on any proposed
topic, inclusive of counterfeit quotations and bogus bibliography. There is
also project of "eating thoughts," as I am inclined to call it: there are special
wafers with mathematical equations written on them; these are to be eaten,
and in due course the equations will be digested, materially assimilated, to
finally reach the brain, without the ordinary toil of thought. Alas, the students are inveterate resisters; they bring up again the bolus, whenever they

can. So much for the great project of reducing mind to digestible material elements.

The irreverent laughter Swift arouses at the counterfeit doubles of scientistic curiosity can be companion to the reverent *studiositas* that is astonished at the wonder of life itself. There is something of the irrepressible festivity of life in affirmative laughter, even when it seems to be only debunking. The return of life, the return to life that is brought to our attention, can be travelled by different paths, some more studious, some in and through life itself, and even its absurdities. Irreverent laughter can serve reverence for life. Serious scientific curiosity counterfeits life and serves death. It cannot laugh, cannot laugh at itself.

Of course, we sometimes say about laughter that nothing seems sacred to it. But this can be said in many senses, some indeed as *incognito* servants of the sacred. Simon Oliver draws us to a crucial question (which joins with the questions of Michael Hanby and David C. Schindler): whether and how desacralizing is itself a condition of the possibility of science, as it has come to be? There is a desacralization that could be seen as non-idolatrous: the difference of the Creator and creation is to be upheld, and both are to be affirmed after their own being and ultimacy. But there is a desacralization that is an assault of creation that "neutralizes" the things, puts out of play the sacred charge of things and the divine communication in and with things. This "neutralization" produces scientistic idolatry. And there can go with it a nihilistic mockery that unbeknownst to itself secretly pays its complement to the sacredness it inverts and abjures.

I find much food for thought in considering whether modern science has always been so married to technology that its curiosity is already framed by the dominion of serviceable disposability *in nuce. In nuce*: yes, since what is in the seed can balloon into a monstrous growth, such as we find in the military-industrial complex and the pharmaceutical industry, the one servant of death in the name of peace, the other said to serve life at war with death. This is not the mustard seed.

Is there a scientism that wants to eat of the tree of life as well as the tree of knowledge, in its manipulation of life? A genetic modification that produces a counterfeit double of genesis, redoing genesis by an undoing? If so, such an undoing redoing of creation would be close to an ultimate counterfeit double. Creation would not be the original child of wonder but a technological changeling. The darnel would be genetically modified to present itself as wheat.

Simon offers a striking reflection on the mutation of creation *ex nihilo* in connection with the vacuum in modern science; it is put very appositely in discussing Robert Boyle, Earl of Cork, and Anglo-Irish, like Swift. I cannot comment on Boyle's sense of humor, and it would be an interesting research to discover if Boyle's projects in the Royal Society might call forth Swiftian laughter. Simon Oliver helps us smile, at the least—and to think. He recounts very appositely the demonstration of the air pump and its exhibition of technological control. The pump "provoked curiosity by producing literally 'nothing'—a void space that could nevertheless extinguish the light of a candle and the life of a bird." This is a nice touch and one is inclined to continue in variation: the air pump, having done its work of undoing, creates a *metaxu* that is airless and in it nothing living can breathe, much less live: the breath of the divine *Ruach* is sucked out, and even though the pump does not curse the darkness, it cannot light a candle . . . and with the *Ruach* sucked out, the living being dissolves in a handful of dust, nothing less, nothing more, nothing anymore. Instead of in-spiring, there is ex-spiring. The ex-spiration of air "creates" a nothing: first breath turned to last breath—spiritually.

I would connect this with Simon's fertile way of bringing this issue together with life in its generative eros: the emergence of the astonishingly new with the birth of the child. The child: not only a new incarnation of astonishment, but the child as offering the renewal of astonishment before this new singular being to those who have served its generation, the parents. Astonishment redoubles in the erotic *metaxu* that generates beyond itself something astonishingly new. The strange actuality and mystery of the "once": "once" a child has been born, it is often felt as if that child has never not been, it is so hard to think the mystery of this life out of the picture, once given to be. This adds another perspective on the child (to those above of Steven Knepper and others): as multiplying wonder in themselves and in those who bring them to be and look on them in love. Wonder-fully said by Simon: "Finally, life always returns, forever repeated in non-identical fashion as spring emerges from winter, and therefore becomes a sacrament of the agapeic character of being's primal gift."

Concerning our flesh I interject a thought about the counterfeit double in its claim to perfect the original, in relation to the infectious vogue for *cosmetic enhancement*. We are adorning animals, and often we refuse to accept ourselves as we are and set out on the quest of beauty reconstructed. The desire to "make-over" one's body as more beautiful can end up ugly, with rictus fixed in an unalterable smile, which is really the inability to smile. The counterfeit double gives us *the perfect smile without smiling*. Smiling itself is a sign of the qualitative "more" of the human being, receiving the sign of hospitality from the other, communicating hospitality to the other, as we see in the spontaneous smile of the young child to the smile of the mother. "Perfected" beauty is the smile without the smile. Perhaps it is an exaggeration but the shock and awe of the cult of the celebrity strikes one, spiritually, a cousin in the same family as the shock and awe of the military industrial complex. Less lethally, photo-shopping is in the business of the counterfeit double, as is the transgender ideology that would "make-over" a man as a woman, a woman as a man. One doubles "as" the other, "as if" one were the other. "Being as" doubles for "being." The equivocity of this doubling is glaringly univocal: being a man without being a man; being a woman without being a woman. The pagan poet Horace come to my mind: *naturam expelles furca, tamen usque recurret* (one may drive out nature with a pitchfork, but she keeps coming back, Epistles i. x. 24).

Simon makes an important point about nature and art, and one might propose a sense of the first art of nature more primal than the derived art of humans. The first art is more nature naturing, while the second art is first natured, and then becomes open to a more original naturing, sometimes called inspiration, and then it is more truly art. But of course, the second art can "lift off," so to say, the first art, and then the will to "create" becomes primarily self-referential, impelling itself forward by a wilful self-insistence that would receive nothing from beyond itself. The first art is closer to the divine creation; and if we not merely lose touch with it but turn against it, revolt against its givenness, in the belief that now finally we are creating ourselves as creators, then we will come to running on empty and the inevitable outcome will be spiritual exhaustion. The breath of the spirit will depart from the living being, and like the air pump will be witness to airless nothing.

It is relevant to remember that the urban experience of many lacks naturing; we are surrounded by human buildings and inevitably we receive back from these mirrors the images of ourselves; but there comes a point where there is nothing of ourselves reflected back to us. If we dwell differently in nature, opening ourselves to an attunement to its naturing, there is a generative otherness that is not reducible to the mirror reflecting us back to ourselves alone. The otherness of nature naturing brings life back to us differently because, in truth, we always are participants and beneficiaries of its gift. The naturing otherness brings us back to ourselves with the promise of new access to original wonder. That as parents we would generate beyond ourselves, that we are offered children themselves wonders and witnesses to renewed wonder, is testimony to the agapeic gift that continues to give and is not depleted but augmented in giving.

XI

Michael Hanby's impressive and thought-provoking **"Being in Control"** echoes a theme in Simon Oliver's reflection. I am grateful for the generosity of his drawing of attention to something singular about my way of thinking and writing, as I am also grateful for the challenge of his questions to me. I hope the first has something to do with trying to be true to the plurivocity of being between system and poetics. The idiotic is not to be slighted, nor is attendance on the companioning power. Being open to wooing, and being wooed, there is a music from the overdeterminate, received into the porosity, impressed on and enjoyed in the *passio*, worded in a *conatus* that endeavors to be true, though not just as asserting itself, but an obedient *conatus*, a listening and, now and then, a hearing endeavor. All of this is out of the time of a certain modernity that lets its epistemic impatience turn into an ontological revolt against the equivocal—a refusal of it when we deem it refuses to cooperate with us. Then we would be the operators, not the co-operators, either with creation as other, or with God.

What Michael Hanby speaks of as "intensive infinity" is present in the plural sense of infinity in my work from *Desire, Dialectic, and Otherness* onwards. There is a resonance of it in the hyperboles of being (especially in *God and the Between*): immanent determinations that exceed immanent determination, overdeterminations in the usage I employ. The hyperboles suggest in finitude manifestations of intensive infinity, though I take God to be overdeterminate to the infinite degree, hence both immanent and transcendent. I am at one in reminders about saturated determinacies, which are themselves overdeterminate in the revelatory senses implied. My point is never to call to account determinacy *tout court* but to come again and again to the gift of its being charged with the overdeterminacy. This is in the spirit of a poet like Hopkins: "the world is charged with the grandeur of God," for "there lives the dearest freshness deep down things."

Michael puts me to the question about the sustainability of the distinction between science and scientism. Perhaps it is best to first address this, and then add some further reflections. There are huge overlaps between what he and I understand, and his thoughts make me confess that perhaps there is a respect in which I am being relatively polite in affirming that distinction relative to modern science. I still think there is sense to making the distinction, but I also take the point Michael is seeking to make about the scientistic impulse being more than a temptation. In some sense, it is already constitutive of the practice of science, and hence any effort to keep open a distinction between science and scientism does raise questions. My response to Michael continues in my response to David C. Schindler.

The somber truth is that the counterfeit self-infinitizing that can over-take scientistic curiosity is multiply manifest, and in a manner that masks the more truthful sources of the counterfeit self-infinitizing in original won-der itself. An element of the counterfeit double is that it often presents itself as a more perfect version of the original from which it secretly derives. We can see this in forms of scientism that both repudiate and claim to perfect earlier and more encompassing understandings of science. This encompass-ing sense must surely be open to metaphysics and theology, and if I am not mistaken Michael makes something of a plea in that direction. It is a plea that too few hear, and yet it has to be given its due. But one must already be in another epistemic space that acknowledges that science in its character-istic determinations of things is an abstraction from the plentitude of the overdeterminacy of given being. (I will say something more about this in response to David C. Schindler.)

I suggested above the possible relevance to the question of scientism of *the parable of the wheat and the darnel growing together*. There is an equivo-cal intertwining of these two in the middle field of being. Suppose these two seeds are not univocally separable as entirely distinct growths. Suppose the seed of the darnel is a kind of mutation of the seed of the wheat. That middle field allows both to grow, grow too as intertwined with each other. The complication of the counterfeit double is that if the growing darnel presents itself as wheat, its mimicry is effective by drawing from the power of the original, the wheat. The difficulty of separating the two by human hand is insuperable, since that hand is itself a carrier of equivocity and alone not wisely steady enough to separate the one from the other. The human hand needs a more ultimate steadying hand.

Can we think of the modern scientific project as the sowing of both seeds of wheat and darnel? (I put aside the question of evil as mimicking the good; as well as the question of metaphysics and theology as sowing their own darnel as wheat). The seeds in the beginning seems almost noth-ing, and nothing alarming ethically or religiously. But it may be that the unfolding of the project begins to show the darnel spreading, and spreading more as exploitative of the wheat, even to the point of claiming to be the true harvest. If now the darnel presents itself as wheat, it also claims to of-fer the fruits for which science is cultivated (a key point for Descartes, for instance), but its offer, in truth, conceals a counterfeit double of nourish-ing, bread-supporting, life-supporting wheat. If we consume this food, we become malnourished. Feeling this effect, we are tempted to eat more of it, taking it to be life-nourishing bread, but we languish the more.

One might take heart: By the fruits one will know them. What if the bitter fruits make it that we also have lost the savor for better fruits? And yet in time the gift of life-giving fruits blessedly applies here too. There is the seed that dying gives life. This is true of original wonder with reference to scientistic curiosity. The more original wonder proves to be an impossible-to-kill reminder of the life-giving seed, and a recall to its generative and sustaining power. Salutary is recall to a more encompassing sense of *scientia* relative to knowing. I take this as a warning not only to scientistic science, but also to forms of metaphysics and theology that, under the influence of the scientistic impulse, are betrayers of the original wonder. In the self-infinitizing of the counterfeit curiosity they put another changeling in the place of the original child. The reborn changeling lives on sufferance. Its truth is its being false to, its infidelity to the intensive infinity.

Scientism thus makes us re-pose the question older than modern science, the question of *scientia* and *sapientia*.[37] This is a question that haunts quite a few of the different essays in their response to my reflections. I take the sapiential way as a guardian angel against a scientific way that would turn away from ethical considerations and religious implications. It could be said that the modern picture of the scientific "project," in its excision of the good in the form of the final causality, and in its will to silo itself as an entirely "autonomous" discipline, has delivered its response to the angel. And yet the companioning power does not vanish.

We see that when we try to silence the companioning power, over the long haul, the sapiential way can itself be counterfeited by a sanctioned culture of "experts," self-certified in their membership of the dominion of serviceable disposability. The ancient issue of the difference and relation of the philosopher and sophist is not irrelevant. The sapiential is somewhat in the hinterlands of my discussion, which tends to strategically zone in on the contraction of curiosity as incarnated in the modern cult of univocity. I think I am in agreement entirely with the thrust of Michael Hanby's reflection, though perhaps not as forthright in stating it. Perhaps there are occasions when one is called to protect the mustard seed (a different image, I know) of truthfulness even in the scientific contraction.

Earlier I implied that we are "successful" because we are "superficial." We cannot jump over our own shadow: the immanent *Ananke*[38] of being true is at work in all our claims to be scientific, even untrue scientistic claims. This inner *Ananke* points back to originating wonder, wonder out-living scientistic curiosity, and wonder the sometimes *incognito* companion of all knowing endeavors in the between. Waking up to this makes one restless for *scientia* in the truer sense here touched on. Alas, the "world"

---

37. Meaning: "knowing" and "wisdom/discernment."
38. Meaning: force/constraint/necessity.

dominantly sets the default terms of answering that restlessness; which is to say it diverts it back to feed again the self-infinitizing circuit of counterfeit curiosity. Revelation from outside this self-circling is the hope for redemption of the true.

The more ultimate sense of sapiential science is hard to state in terms of "science," since the latter term in the more modern sense overhangs all discussion. There is a theme here that lines up with, among others, Simon Oliver, and pointing to a similar engagement in David C. Schindler: how we move from love to control; and perhaps too how love returns. The theme of the connection of knowing and power, especially in the modern project, is frequently stated, and rightly so. But we ought to remember that the project is enacted *without the good* in the earlier sense, as the bearer of metaphysical, ethical, and theological weight. I mean the extrusion of the good as final cause from the modern scientific scheme of intelligibility, and this from the outset. This is the good not in our moralistic sense, but in an ontological sense, indeed theological sense, that transforms our ethical sense.

Without the good what is there becomes worthless being, purposeless process, ongoingness without point—unless putatively we give it purpose, and step forth as the good of the whole. But if there is to be a universal application of the divorce of being and good, a divorce demanded precisely because one cannot univocally mathematize the good, the divorce has to be applied to us also. We, as parts of the worthless whole, are ourselves worthless in the horizon of the whole, and our proud claims to "create" values themselves come to nothing. Creating value amounts to counterfeiting value. "Value" is backed by nothing but our assertive say-so. This means that the extrusion of the good means that the seed of nihilistic scientism is there in the root of the scientific project. Sapiential science would concern a knowing of being which is good—knowing the good of the "to be," and of our "to be" as good and to be good. Paradoxically, all scientisms are carriers of sometimes secret, sometimes proclaimed candidates for what is to be as good. They rarely come clean about what they carry or proclaim. They are counterfeits of *sapientia*.

One finds all but nothing reflecting this matter in any of the scientisms. There is a dearth of mindfulness mirroring the dearth of astonishment, and especially bearing on the good that is to be sought, striven for, brought to be by us, supposedly. The desire to know is already fallen at the outset if the desire of the good is not granted. The good in the sense at issue is itself counterfeited in moralistic terms. The love of being and being true becomes a loveless love of knowledge: loveless love—showing the face of the equivocity; revealing a counterfeit double of desire. Such desire does not seek the beloved and its good but control of it.

Loveless love of knowing: this is the poisoned equivocity out of which thrust the seeds of the infernal darnel. The ecological devastation in the middle field of being is not only of nature around us. We too become loveless weeds, self-growing in the garden of creation we turn into a waste. We become the worm in the apple condemning the apple for being worm-eaten.

XII

There are many excellent things in **David C. Schindler's** reflection **"Wondering about the Science/Scientism Distinction."** There is, for instance, his emphasis on the ontological stress of wonder—it is not just a subjective feeling of wow but has to do with the revealing of the things in their own being to us and not just simply for us. It awakens us to being in the midst of things in a manner that calls into question any simple distinction between science and scientism, in a manner strongly resonating with Michael Hanby. My response to Michael (as well as some things I say in response to Simon Oliver) can be consulted as also trying to address this merited question from David. He proposes a more holistic sense of things, which brings out that science, granted in its positive meaning, is itself an abstraction from the rich overdeterminacy of being and not coincident with its fullness, and that by implication scientism is even more an abstraction, since it seems asleep to this fact about science itself. It adds insult to injury, so to say, by generalizing itself to the whole, producing in the process a counterfeit whole, all the while giving itself philosophically thoughtless credit about being on a par with the true whole. These extremes of subjectivizing the origin and totalizing the end, and each in a counterfeiting mode, are very important points that are made.

Though I see here an overlap with Michael Hanby, there seems more the possibility of opening up some chink of difference between science and scientism in David's way of putting the matter. I do not want to deny this opening in Michael, nor here exaggerate the degree of worry about ontological infidelity that lies at the heart of David's thoughts also. I think both would agree that the more encompassing sense of *scientia*, if not antagonistic to the sapiential, is to be granted, and is implied as a needed horizon to think the significance of wonder, astonishment, curiosity, and scientism.

I would like to speak of an "open whole," or, theologically, of creation as an immanent whole open to God as other. In *God and the Between* chapter 11 I speak of "God(s) of the Whole: On Pantheism and Panentheism," while in chapter 12 I speak of "God beyond the Whole: On the Theistic God of Creation." I take issue with the absolutization of the "immanent frame," as Charles Taylor speaks of this. That absolutization comes down to a metaphysical postulate, a postulatory finitism, masquerading as anti-metaphysical, or non-metaphysical, or post-metaphysical. These latter are counterfeit doubles of some secret metaphysics, unacknowledged to others, unknown to itself. If original astonishment is opened to the whole, it is also opened beyond the whole, and finds itself perforated by communication beyond any postulated immanent frame.

Present in David's discussion is a recognition that the subjectification of wonder has intimate repercussion of the negative self-determination of doubt, leading from Descartes to Hegel, and beyond. In Hegel claims of the triumph of thought through self-relating negativity are of a piece with his claim to be true to the whole, and his claim that the true is the whole. All of this is of a piece with David's emphasis on a needed holism at the end. I have argued that self-relating negativity cannot generate anything truly affirmative, if it does not call on an unacknowledged endowment of being and mind, not to be described in terms of the negative. Without that acknowledgement the seed of a nihilism is sown in the negativity, so understood.

Are there forms of holism that can feed the growing of false senses of the whole, indeed the construction of counterfeits of the whole? I see something of this in the case of Hegel. Hegel speaks true when he says the true is the whole. But the way the true and the whole are understood by him lead via negativity to an all-inclusive logic of self-determining thought, relative to which there is nothing finally other. David is an adept in engagement with Hegel, perhaps more of a gentleman in addressing him than I am. Critics of my view of Hegel forget that when I bid Hegel adieu, adieu is not alone a leave-taking but also a blessing. One has to be able to take ones leave and yet offer blessings to the one from whom one departs. Hegel can be seen as a thinker of the between but how he articulates the between is the issue, and dialectic for him epitomizes and is governed by a logic of self-determination that converts the negative reason of dialectic into speculative reason. This claims to have the measure of, indeed exceed the infinitely qualitative difference of the divine. This last way of speaking joins with Kierkegaard, and on this Kierkegaard is not wrong. Kierkegaard converts dialectic to the beyond of dialectic. This beyond is not truly understood by Hegel's speculative reason. The truth of dialectic is metaxological. There is a beyond, a transcendence as other that opens every immanent closure of any holism on itself. Holism cannot be the whole, so to say, cannot be the whole of betweening, whether of creation, of the human creature, of the divine creator.

I do spend some time hovering around Hegel's "speculative scientism," as I put it. The issue of the counterfeit double does not only apply to more ordinarily recognized forms of scientism, which we associate with a particular science making claims to be the one true key to the whole. The *postestas cla-vium* can be claimed by philosophical dialectic also. The point applies with some subtlety to Hegel in that he is clearly a critic of the finite limits of the particular sciences; he is a critic of the contraction of the scientistic form of curiosity, whose self-infinitizing takes the form of the "bad infinite." There is an infinity to thinking, he holds, that exceeds the terms of such an approach, and we require due appreciation of art, and due respect for religion to do justice to what is ultimate, for him absolute spirit.

So far so good, in this display of discernment for more than the merely finite infinitized. And yet the kind of holism Hegel gives us, when all the complexities are considered, is absolutely self-mediating, self-circling, with nothing finally other or beyond it. The gap between our infinity *in potentia* and God's infinity *in actu* is dialectically closed. Hegel's holism is that of Spinozistic Substance raised to Subject, understood as absolute *Geist*. One might say that in granting the infinite restlessness of the desire to know, this restlessness is put to rest by circling around itself and circling in itself, in the enacting of a divinity not other to the thought that thinks itself. This teleology of wonder might grant: astonishment in the beginning, but in the end not determinate science, but self-determining philosophy as achieved *Wissenschaft*, or Science, overcoming and dispelling the initially indeterminate wonder. Speculative *scientia* overtakes philosophical and religious *sapientia*. My main point here would be the closure of the immanent circle in this version of holism. Looked at in a certain way, it is the unremitting immanence of this response that puts it in the same family as the unremitting immanences of other varieties of putatively more finite and false forms of scientisms, associated with one or other of the particular sciences. After all, it is not difficult to see a variation of the Hegelian teleology of history, culminating in his philosophy, in the "little leagues" of scientisms, for instance, in the Comtean teleology of positive sciences and its contemporary supersession of religion and metaphysics. Hegel's speculative scientism is big-league immanent holism.

The "little leagues" draw more immediate attention but the "big leagues" have historical stamina that endures, endures even in their lessers. What deeply worries me about this is that the seduction of this speculative counterfeit is less likely to see the meaning of wonder as perforating immanent holisms. Perforating as opening it from within, relative to source of otherness at work in immanence; perforating as opening from without, in relation to the abiding mystery of creation as other; and perforating it from above, as ever tender to the communication of divine transcendence from beyond every closure of immanence on itself. I think many works of David C. Schindler move in the element of bringing back to memorial mindfulness these different perforations of any holism closed in on itself in immanence. It may sound too grand, but in a way the perforation, the opening and reopening of the porosity, is going on all the time, as much in the humble elements of everyday life and love as in the ambitious projects of the determinate sciences, as in the inspiration of the artists, the prayers of the faithful, and the perplexed rumination of genuine philosophers.

David rightly and provocatively poses the question of what "What would a science that understood itself as ontologically and methodologically guided by the mystery of being as overfullness *look* like? Is such a science anywhere in evidence today?" I cannot give a straight yes or no in answer, and am more inclined to no than yes, though I do not doubt there are practicing scientist in whom original astonishment longs unknowingly for such a science. Metaphysics, in bad odor, and unfairly and uncomprehendingly so, would be a precondition of reformed science. David does ask how figures like Bernardino Telesio, Francis Bacon, Galileo, Boyle, Newton, and so forth, might look as measured by the standard of "astonishment." Not very well, I am inclined to reply. I would suspect many of them as very much in flight from astonishment, in the original sense I mean. Might they be worked through differently and understood otherwise? A good suggestion and not a rhetorical question, on David's part. A further enlargement of the question of "astonishment," as I present it, is perhaps in mind. And he does pose the "massive" question of current practitioners who are or might be companions in the work and reworking. I would like to meet them. They are probably pariahs, and not able to get "grants" for their "projects." But then if they are there, they will not speak the language of "projects."

David is right that I do not say much about what genuine science is. My focus is on the enabling ontological and epistemic conditions that make science possible, but not only science, and how any view of these that overlooks the overdeterminacy of being and our porosity to it in original astonishment is a cognitive bow shooting arrows in a skewed direction. His question is not inappropriate but it would require a differently focused exploration, and it is as massive a question as the question about contemporary practitioners of science in the vineyard of truth. My hesitation here has to do with the way modern science has taken over the whole field of science; and my wonder if perhaps it is best not to call truer science "science" in this sense. And not simply call it philosophical science either, since this would smack too much of the efforts to mimic science by philosophy in modernity, even though modern science itself grows out of a dream of philosophy. One has to have the memory of the holistic sense of philosophy not only as science but also as sapiential, and for the latter we need philosophical finesse not geometrical science. The more reformed encompassing mindfulness I would put down to a *scientia* as metanoetically reformed sapientially—reformed along metaxological lines. In one sense, this is a dream; in another sense, my own attempts at metaxological philosophizing are essays in the direction of wording something of the dream.

Perhaps the ventures of science in Hegel and Schelling were among the last great efforts to rethink the meaning of science in the holistic direction. I appreciate this. But as I suggested above, speculative scientism is something to be navigated also. My own work in reforming and transforming Hegelian dialectic in a metaxological direction means that science itself also cannot absolutize itself. And while Schelling is more open to the "that it is" of being and the givenness of revelation as preceding and exceeding the self-enclosure of thought, there still is needed much of metaxological rethinking, and not least in relation to his perduring Spinozistic tilt. If there is a new formation or reformation, I recommend practices of thought "between system and poetics." Given the contemporary usage of the word "science," this would be both an unrecognized and unacknowledged usage of "science." And this is to say nothing of the too often accessory nature of science and its "projects" to military and pharmaceutical industrial complexes. A metaxological science would acknowledge the beyond of science, witnessing to the originating wonder that is beyond the determinate curiosities of the sciences, and in excess of them finally. The sapiential will always bring back ethical and religious practices, something not straightforwardly scientific, even granting the distinction of geometry and finesse. Many of the responses to my contribution recall us to the sapiential. This always calls on *the betweening of the witness*. I only cite Socrates and his living dialectic as a philosophical exemplar.

XIII

I found it a kind of graced serendipity to have received and read last **Isidoros Katsos's** offering on **"Basil and Desmond on Wonder and the Astonishing Return of Christian Metaphysics."** Paradoxically so, since it opens by calling attention to a new beginning—baptism and indeed the baptism of Christ, and with hope of a new beginning also for the companionship of Christianity and metaphysics. I call attention to this to begin and will end with this beginning. One is made to ask: Can the children of wonder be baptized by the Child of Wonder?

We began above with Spike Bucklow's reflection on painting the Virgin's robe, and now too the Virgin returns as a Mother in the unique icon *Megas ei Kyrie* ("Great are you, O Lord"), painted on wood by Ioannis

Kornaros in 1770. The name of the icon is not that of a person but a doxological praise centred on the motif of Christ's baptism. That this is celebrated in the East on the feast of Theophany (January 6) brings us to the time of wonder of the intersection of time and eternity. As Isodoros points out, attention shows the central motif to be placed "*between* heaven (the Trinity and angels) and earth (Mary with Christ Child, Adam and Eve), followed by the descent of Christ in the underworld depicted at the bottom." To the right and left of this between space we behold fifty-seven selected themes from the Old and the New Testament. That they are all connected through running waters suggests the great purifying, recreating, and engendering power of baptizing. Wonder strikes again: the child, the adult being baptized, the Son of God being baptized.

In its own way, this essay is a hymn of praise to creation, creation as given with its abiding saturations of the signs of God, and a new creation greeting in jubilant hope as to what was, is, and is to come. My response, perhaps appropriately, will be as the echoing back and forth of praise to praise.

I am illuminated and edified to be instructed on the saturation of these texts and particularly the writings of Basil with the word and with the words of amazement and astonishment. I can only acknowledge my gratitude for the nudge towards attentive reading of Basil. I think here we are confronted with living liturgy of the hyperboles of being, gloriously enacted in the divine dramatics of the sacraments and most especially the eucharistic meal. We speak of the Last Supper, but I will mention below a later meal mentioned in the scripture, which is breakfast at the break of day. *Gus am bris an lá* (until the break of day):[39] this is a prayer in Gaelic found on gravestones, expecting in hope the eschatological life to come and the meeting again, the betweening with the resurrected dead, every shadow of the night of death scattered.

To say something first of old metaphysics: I am instructed on the great importance of the astonishing in the works of Aristotle as reworded in this reflection. I will confess that my soul tilts mostly more Plato-wise than Aristotle-wise. I confess that in the life to come I would prefer to talk to Plato first before talking to Aristotle. Among my reasons: Plato's first identification of *thaumos* as the pathos of the philosopher. I had thought of this as not quite fully understood by Aristotle because of what seems a strong orientation to intelligibility as tied to determinacy—to be intelligible is to be a this somewhat, a *tode ti*.

---

39. Alasdair McIntyre uses this as an epigraph for his *After Virtue*, honoring in particular an uncle, an epigraph puzzling to some.

We are familiar with the often-cited remark: "All men desire to know" (*Metaphysics* 982b11ff.). Aristotle does see the connection of marveling and astonishment when he reminds us of the affiliation of myth and metaphysics, as well as the delight in the senses. I take him to stress that our desire to know is a drive to determinate intelligibility. When we attain this, the wonder launching the quest is dissolved. If the end of Aristotle's wonder seems to be a determinate *logos* of a determinate somewhat, a *tode ti*, this would suggest the dissolution of wonder, not its deepening. At least so it seems when Aristotle names *geometry* to illustrate the teleological thrust of our desire to know (*Metaphysics* 983a13ff.). One might see geometry as representing determinate cognition of a determinate somewhat whose attainment solves the problem but also leaves behind the wonder.

Present throughout the philosophical tradition, and exemplified in determining science is something like such a teleology of knowing, moving from indeterminate wonder through determinations of curiosity to a determinate solution dispersing the original wonder. This does not do enough justice to the *pathos*, in the origin and in the end too; not enough justice to the porosity of being and the *passio essendi*, that are the more precedent enabling sources of the desire to know itself. The outcome is that the desire to know becomes a drive impelled by a cognizing *conatus*, carried teleologically to dissolving puzzles in consummate, determinate solutions.

Still, there are further citations by Isodoros of Aristotle that offer a redress to the balance and indeed show the companioning of Plato in Aristotle himself. I refer to the complication of the matter when Aristotle speaks of the *divine knowing* (in which we sometimes participate in *theōria*) as wonderful (*thaumaston*); indeed it is even more wonderful still (*thaumasiōteron*) if divine knowing exceeds ours, as it does (*Metaphysics* 1072b24–27). This is important. Such a divine knowledge is not the determinate cognition of geometry but of the thought that thinks itself. I only recall that Hegel refers to this famous passage at the culmination of his *Encylopaedia of the Philosophical Sciences*: in Greek too, not in German. The living and eternal enjoyment of such self-knowing is the possession of the self-thinking God. We are beyond the determinate cognition of a determinate somewhat; we are in the freer space of *self-determining knowing*. My question is whether there is also *more* in *our* access to wonder than can be exhausted by self-determining knowing. The question extends to the knowing of God, whether there is even more also to the God of agapeic mindfulness, the God who creates and comes among us and offers himself in birth, and growth to redeeming death, and to redeem from death.

Apart from God, I confess a fondness for the way Plato puts wonder in *Theaetetus* (155d3–4) where *thaumazein* is named as the *pathos* of the philosopher. *Pathos:* there is a patience, a primal receptivity. This is not the self-activating knowing such as we have come to expect from Kant and his successors in German idealism, as well as in varieties of the constructivist epistemology in different contemporary inheritors of this Kantian stress. There is a pathos more primal than activity, a patience of the soul before any self-activity. In modernity patience has often been relegated to a servile passivity supposedly beneath the high dignity of human power as self-activating, as self-determining. But no one can self-activate themselves into wonder—except perhaps into a counterfeit curiosity, an "as if" wonder, a pretence of wonder. Wonder comes, or it does not come. As has been repeated, we are *struck* into wonder. "Being struck" is beyond our self-determination. We cannot "project" ourselves into "being struck." It comes to us from beyond ourselves. Like "beholding from" it comes in the communication of an intimate strangeness that makes us porous to what before us is enigmatic and mysterious.

Speaking Greek, I would also like to acknowledge and find reinforced here the *iconic* nature of the issues at stake. Both the negativity of speculative thinking à la Hegel's self-determining thought and the will to determinacy involved in the scientistic configuration of curiosity have too little patience, if any, for the iconic dimension, in its precedence and exceedance of our mortal rationality. Thus Hegel's move to conceptually sublate the religious representation; thus scientism's impatience with the poetics of the sacred and the hyperboles of thought of mindful metaphysics. Again one invokes Comte: religion, metaphysics, positive science, . . . but where is the hymn to the Creator in creation, the icon of the aesthetic and trans-aesthetic God?

I am brought back to a special point of reference in my own thinking in respect of any metaphysics of image and original. I mean Plato's *Timaeus:* itself a likely, that is to say iconic, story, not a *Wissenschaft* or system; and gloriously announcing that "it is altogether necessary that the entirety of the cosmos," itself a kind of aesthetic god, "is a certain icon." This is the striking sentence, just noted in translation, now in the original Greek: "πᾶσα ἀνάγκη τόνδε τὸν κόσμον εἰκόνα τινὸς εἶναι (*Timaeus* 29B)."

Has this something to say to metaphysics still living, or coming to life again? In my own essays in metaxological metaphysics this Platonic wording has provided a locus of aesthetic, metaphysical, and theological rumination.[40] The metaphysics of image and original inherent in the necessity of the entire cosmos being an icon supports metaxological metaphysics and indeed the philosophy of God that goes with it. The hyperboles of being reflect this as determinations in immanence that yet exceed immanent

---

40. This utterance is the epigraph to my *Art, Origins, Otherness*.

determination. There is a double, swivelling motion on the threshold of immanence that reveals immanence in its plenitude and yet turns beyond and
gives saturated signs of what exceeds immanence, just in its being given to
be at all, and its being given to be as beautiful and good.

To return to the icon and the eucharistic meal, we often forgot that
the Last Supper was not the last meal Christ shared with his disciples. There
was the meal of the stranger, the mysterious companion who walked with
the disciples fleeing from Jerusalem and with whom they broke bread in
Emmaus. There was the last meal mentioned: homely breakfast being prepared on the seashore, and perhaps most fitting that it was the breaking of
a fast, after the desolation of the death of Christ, and dawning of a new day.
Dawning in the midst of the elemental things of creation, sea, sky, land,
Christ—"beholding from" a metaxological fourfold that gives a different
ontological, theological constellation of new being, other than the Heideggerian fourfold (*das Geviert*).

I found the graced serendipity of Isodoros's thoughts deepened in that
the invocation of the icon brought me back to words and reflections recently
delivered at a meeting on "The Future of Christian Thinking."[41] My title
was: "Christening Philosophy." I have sometimes said that if one is stuck
in a rut one has to rock back to release forward motion, and sometimes
the old wheels are superior to new wheels, old theological, metaphysical
wheels rather than post-metaphysical and non-theological or a-theological
wheels, new as new but not always good for rolling. I have tried to speak of
a *companioning* relation between philosophy and religion/theology. There
was philosophy before Christianity; there was the exemplary Christianiz-
ing of philosophy with thinkers like Augustine and Aquinas; there was the
growing separation of Christianity and philosophy in the modern period,
sourced partly in philosophy's assertion of its autonomy, though there are
other sources too; then there is a de-Christianizing of philosophy, following
by a post-Christian form, the philosophical mind blank of any memory of
the companionship held out to it by Christianity.

My title, "Christening Philosophy" was with a view to a companioning
of Christianity and philosophy to come after that, trying to take some measure of the de-Christianized and post-Christian world of the West. Christening is raising new life to a higher power of divine newness of life. It is not
done through our own power, through our own self-determination. There
is our being placed again in the primal porosity between the divine and the
creature; this is to be in a sacred betweening where the pouring of water not
only purifies but also offers the life-promising matrix of new life. The waters

41. At St. Patrick's Pontifical University, Maynooth, Ireland, 27–30 April 2022.

of Christening are not the waters of *tohu wa bohu*,[42] nor are they the waters of the flood. They are the chaste tokens/signs of pure flowing, pure porosity

Isodoros's meditation here, as it were, christens my own essay on the dearth of astonishment. Given the baptismal theme, I want to refer to some of closing lines from "Christening Philosophy." I was struck by a community of spirit between Isodoros Katsos and what I was trying to say. I am asking about the de-Christianizing of philosophy after the death of God:

"What would it be like to walk with the stranger on the way to Emmaus? Philosophy and religion as companions, but along the way a companion joins them and walks with them? How walk with the still unrecognized Christ, Christ not yet for us Christened, we as philosophers hurrying out of, away from Jerusalem? Along the way the stranger is not Christened, but at the breaking of bread, at companioning (*cum-panis*), there is the astonished dawning of who the Stranger is, as he vanishes. Amazement at his understanding, his wording of things, his wording of the things one had heard before in the old writings but now newly illuminated, newly oldly making astonishing sense: would not all this throw our own philosophical wording of things into disarray? Could one continue wording the things in the old philosophical way? Or would the way no longer be either old nor new, but both and neither, intimate to each and yet beyond both? What new wording would or could come to one, be offered to one? Would ones claim to philosophical "autonomy" seem unreal, seem abstract, seem untrue, an inability to listen, a refusal to listen, a disability of hearing (*oboedire*: give ear)? On the way to: listening, wording, differently, unrecognizable in the univocal, even dialectical vocabularies, perhaps plurivocal? Speaking in tongues, and yet entirely philosophically?

After the dawning, the disciples return to Jerusalem by two, as previously they were sent on mission by two. By two: *betweonum, bij twee*, betweening. As if companioning Christ, the one who breaks bread with us, has now offered us to be christened to a second power, called into resurrected betweening, and to a hyperbolic degree. The full Pentecostal *parousia* for the community is to come, and perhaps we pray, *gus am bris an lá*, until the break of day. Perhaps the disciples remember there was before such a theophany, when Jesus was baptized in the waters of the Jordan. Christ himself christened: the christening of Christ seems odd and yet is it not companioning raised to the hyperbolic degree? The agapeic generosity of the divine sharing the sacred sign of purgation? Not only purgation of the sacred porosity, and elevation, it is also immersion in the waters of the flood,

---

42. Meaning: "formless and empty." Referencing the primeval waters in the first Genesis creation story (Gen 1:2).

and going into the belly of the beast, the leviathan, and coming up and out again on the banks of the river, sovereign over the *tohu wa-bohu*, on the side of the flow of time and on the surface of the earth, where space itself is sanctified, and from whence will be harvested bread and wine; and even when we are at sea, we are to behold on the shore, behold from the shore, the resurrected companion, the resurrected Child of Wonder, and not a gnostic changeling, preparing breakfast for friends at break of day, after a long night of toilsome fishing, in the mysterious waters of the infinite sea.

Bibliography

Bucklow, Spike. *The Alchemy of Paint: Art, Science and Secrets from the Middle Ages.* London: Boyars, 2009.

Desmond, William. *Art, Origins, Otherness: Between Philosophy and Art.* Albany, NY: SUNY, 2003.

———. *Being and the Between.* Albany, NY: SUNY Press, 1995.

———. *Ethics and the Between.* Albany, NY: SUNY Press, 2003.

———. *God and the Between.* Oxford: Blackwell, 2008.

———. *Godsends: From Default Atheism to the Surprise of Revelation.* Notre Dame, IN: University of Notre Dame Press, 2021.

———. *Is There a Sabbath for Thought? Between Religion and Philosophy.* New York: Fordham University Press, 2005.

———. *Philosophy and Its Others: Ways of Being and Mind.* Albany, NY: SUNY Press, 1990.

Duns, Ryan G. SJ., *Spiritual Exercises for a Secular Age: Desmond and the Quest for God.* Notre Dame, IN: University of Notre Dame Press, 2020.

Hart, Ray. *God Being Nothing: Towards a Theogony.* Chicago: University of Chicago Press, 2016.

Heidegger, Martin. *Country Path Conversations.* Translated by Bret Davis. Bloomington, IN: Indiana University Press, 2010.

———. *Pathmarks.* Edited by William McNeill. Cambridge: Cambridge University Press, 1998.

———. *The Question Concerning Technology and Other Essays.* Translated by William Lovitt. New York: Harper and Row, 1982.

Knepper, Steven E. *Wonder Strikes: Approaching Aesthetics and Literature with William Desmond.* Albany, NY: SUNY Press, 2022.

Kuhn, Thomas S. *The Structure of Scientific Revolutions.* Chicago: University of Chicago Press, 1970.

McIntyre, Alasdair. *After Virtue: A Study in Moral Theory.* Notre Dame, IN: University of Notre Dame Press, 1981.

Tyson, Paul. *Kierkegaard's Theological Sociology: Prophetic Fire for the Present Age.* Eugene, OR: Cascade, 2019.

Printed in Australia
Ingram Content Group Australia Pty Ltd
AUHW020919210923
384003AU00014B/76